A Cultural-Historical Study of Children Learning Science

Cultural Studies of Science Education
Volume 11

The series is unique in focusing on the publication of scholarly works that employ social and cultural perspectives as foundations for research and other scholarly activities in the three fields implied in its title: science education, education, and social studies of science.

The aim of the series is to establish bridges to related fields, such as those concerned with the social studies of science, public understanding of science, science/technology and human values, or science and literacy. *Cultural Studies of Science Education*, the book series explicitly aims at establishing such bridges and at building new communities at the interface of currently distinct discourses. In this way, the current almost exclusive focus on science education on school learning would be expanded becoming instead a focus on science education as a cultural, cross-age, cross-class, and cross-disciplinary phenomenon.

The book series is conceived as a parallel to the journal *Cultural Studies of Science Education*, opening up avenues for publishing works that do not fit into the limited amount of space and topics that can be covered within the same text.

More information about this series at http://www.springer.com/series/8286

Marilyn Fleer • Niklas Pramling

A Cultural-Historical Study of Children Learning Science

Foregrounding Affective Imagination in Play-based Settings

Springer

Marilyn Fleer
Monash University
Frankston, VIC, Australia

Niklas Pramling
University of Gothenburg
Gothenburg, Sweden

ISSN 1879-7229 ISSN 1879-7237 (electronic)
ISBN 978-94-017-9369-8 ISBN 978-94-017-9370-4 (eBook)
DOI 10.1007/978-94-017-9370-4
Springer Dordrecht Heidelberg New York London

Library of Congress Control Number: 2014947233

Printed on acid-free paper

Springer is part of Springer Science+Business Media (www.springer.com)

Foreword

Personally and as an educator, I have always found it helpful to think of Science as a human invention in which I could share. So it was pleasing to find the authors of this book unequivocally describe Science as a human invention. Very often in school science, recognising or recording a natural phenomenon is confused with its science. Phrases like "discovery learning" and "guided discovery" commonly confuse this recognition of a natural phenomenon with its subsequent description and explanation in Science. Making the distinction clear helps science educators to explore and teachers to employ pedagogical processes that move a learner from recognition to scientific understanding. This book is about a particular view of how early childhood education can facilitate these processes.

Describing Science as a great human invention locates it alongside those other great human inventions like Art, Music, God and Technology, each of which has an immanent (personal) sense and a transcendental (beyond the personal) sense. The cultural-historical foundation for early childhood science education that Marilyn Fleer and Niklas Pramling, the authors of this book, espouse matches these two senses, and they skilfully interrelate them.

In the late 1960s, soon after my career in science education began, I visited the Elementary Science Study (ESS) project in Massachusetts, a more interesting one of the many science curriculum projects then being developed on both sides of the Atlantic. A riveting comment by its director, Robert Hein, still comes easily to mind. "Science should be the easiest subject to teach in the early years of schooling, because it only requires the abilities to see and to describe in talk what you are seeing, and the great majority of children entering school have these abilities already." The materials for this project (and its counterpart in England, Nuffield Primary Science) suggested that teachers should encourage their young students to engage with a natural phenomenon and then develop the lesson from the questions the students asked. There were many reasons why these projects failed, but three come to mind in relation to this book.

I have already referred to one, namely these projects' mantra of "discovery learning" insufficiently recognised the invented nature of the science that described and explained these natural phenomena. The second is the assumption the projects made that curiosity about nature is an innate quality we all share. I do not want here to be definitive about such a deep issue. We would not have Science at all if there had not been persons, since the dawn of humanity, who wondered in awe, and then in curiosity about nature. It is enough to say that these responses were not equally shared when the ESS and Nuffield projects' materials were tried in schools. Only some of the students were interested in the engagement phase and fewer still came up with investigable questions. Furthermore, many of the teachers in these early school years were not interested in or inspired by these phenomena.

This widespread lack of interest in the natural world is something I have had to become more and more aware of in the ensuing years. When I began to think about the issue of Science for All in the 1980s, I had to come to terms with the fact that most of my peers in school did not share the same fascination that I had with Science's account of natural phenomena. Otherwise, they too would have chosen to study the sciences in senior secondary school and go on with them after school. So I can accept Fleer and Pramling's premise that wonder in the face of nature is not innate but needs help to be acquired. This acquisition needs to be in both meanings of the word – the sense of sheer wonder (awe) of a phenomenon and the sense of wonder (starting to question) about the phenomenon. This book deals particularly with this second sense of wonder about the natural world (which is called curiosity) not being innately shared and importantly attends to how one common source, the experiences of pre-school education, can and should contribute to its learning.

The third reason relates to how rapidly formal teaching and learning about a natural phenomenon should move towards Science's account of it. Too often, the school science curriculum has tried to rush this movement. I remember going into a bakery in a small country town in Victoria and being told by the baker that a local teacher had recently brought her prep year students (5-year-olds) on a visit. The baker told me he had been surprised that the focus of the visit was not on how bread and other products were made but on how any heavy objects he used were lifted and moved. Equally curious, I went to the school and talked to the teacher. She showed me a range of "heavy" objects that had been put in the playground for the students to explore how they might move them. Furthermore, she explained that her group of students had also visited the local garage and the small supermarket to see how their heavy objects were moved. Through these teaching experiences, the students had come up with a number of ways to move their set of objects. Each of these ways embodied in the students a set of experiences that would stand them in good stead in subsequent years at school to engage with the theoretical concepts and principles that Physics, as a science, has used to account for the movement of heavy material objects.

About the same time, I came across the phrase "preconceptual learning" for the first time. It was coined by the team in the New Zealand Children's Science Project at the University of Waikato, which did so much in the 1980s to inform us all about the ideas and supporting rationales that pre-schoolers and young learners commonly hold about scientific phenomena and about the words like "force", "floating and sinking", "light and dark", "life" and "hot and cold". "Preconceptual learning" was, this team argued, a necessary step that had been very largely overlooked in school science. In other words, students were being expected to learn Lever Laws, Archimedes Principle, Laws of Reflection and of Motion, the Characteristics of Living Things and Chemical Reactions before they had had any significant experience of the phenomena these scientific concepts and principles had been invented to describe and explain. This book reinforces the positive role the pre-school years can play in providing this rich base of experiential learning.

I commend the authors for presenting so clearly the potential that the context of pre-school education has for the beginnings of science learning and commend it to teacher educators and their students – preservice teachers – to take up the opportunities this book offers to them.

Monash University, Clayton, VIC, Australia Peter Fensham
Queensland University of Technology, Brisbane,
QLD, Australia
16 August 2013

Preface

In this book, we move beyond the traditional constructivist and social-constructivist view of learning and development in science. We argue that science as a body of knowledge is something that humans have constructed (historically) and reconstructed (contemporarily) to meet human needs. As such, this human invention acts as an evolving cultural tool for supporting and helping to understand everyday life. We draw upon cultural-historical theory in order to theorise early childhood science education in relation to our current globalised education contexts. We do not seek to make cultural comparisons, as are found in cross-cultural research. But, rather, we seek to better understand the many ways that science concepts are learned by very young children.

The book is designed for researchers and educators interested in a theoretical discussion of the cultural-historical foundation for early childhood science education. In a book of this kind, it is important to examine the contemporary theories of learning and development within the general field of early childhood education. A theoretical examination of this kind allows for the foundational pedagogical context of the young learner to be interrogated. Through this kind of analysis, it is possible to examine play-based contexts in relation to opportunities for scientific conceptual development of young children. With this approach in mind, and with the empirical literature relevant to early childhood education examined, it is possible to introduce a more relevant approach to the teaching of science and for the development of young children's scientific thinking. In this book, we specifically present a pedagogical model for introducing scientific concepts to young children in play-based settings.

Recommended Citation

Professors Fleer and Pramling, editors of this book, jointly conceptualised the content of this book, even though specific chapters have been attributed to specific authors due to reporting requirements to funding bodies (Marilyn Fleer – Chapters 1–5 and 13; Marilyn Fleer and Niklas Pramling – Chapter 6; and Niklas Pramling – Chapters 7–12). As such, we recommend that the content of the book be cited as follows:

Fleer, M., & Pramling, N. (2014). *A cultural-historical study of children learning science: Foregrounding affective imagination in play-based settings*. Dordrecht, The Netherlands: Springer.

Acknowledgements

Throughout the chapters of this book have been presented data and analyses funded by a series of grants either won or that framed the work of both authors. The authors would like to acknowledge the funding bodies for their contributions to supporting research and scholarship in early childhood.

Fleer would like to specifically acknowledge the Australian Research Council (ARC) Discovery Grant that provided the resources for enabling the blue skies research to be undertaken (ARC 2005–2007; 2010–2013; 2013–2015). Fleer is particularly appreciative of the research assistance and leadership provided by Sue March (field leader) and a volunteer student research team which included Feiyan Chen, Judith Gomes, Shukla Sikder and Devi Sukmawati and her colleague Marie Hammer, who supported the professional development programme that underpinned one of the three ARC grants received by Fleer. She would also like to acknowledge Feiyan Chen who provided valuable support with sourcing articles, particularly from the region of China, and who did reference checks and some formatting of the manuscript. Special acknowledgement goes to Ng Geok Eng, who provided the photographs and slowmation example illustrated in Chap. 1, and the work of Avis Ridgway, who was the field researcher for a study reported later in the book. Pramling's part of the research reported in this book is based on his work as a Research Assistant with the Swedish Research Council (2008–2011). Together, the funding bodies and the subsequent research assistance provided invaluable support for the foundational work reported in this book.

Contents

About the Authors

Marilyn Fleer, Ph.D., M.A., M.Ed. (first class Hons), B.E.C.E. Professor Marilyn Fleer holds the Foundation Chair of Early Childhood Education at Monash University, Australia, and is the President of the International Society for Cultural Activity Research (ISCAR). Her research interests focus on early years learning and development, with special attention on pedagogy, culture, science and technology. She can be contacted at marilyn.fleer@monash.edu

Niklas Pramling, Ph.D. Professor Niklas Pramling has a background in studies in psychology and literature history and a PhD in Educational Research. His main field of interest is teacher–child communication in regard to various domains of knowledge, such as science and the arts (particularly music). He is a member of the Linnaeus Centre for Research on Learning, Interaction and Mediated Communication in Contemporary Society (LinCS) funded by the Swedish Research Council. He can be contacted at niklas.pramling@ped.gu.se

Part I
Theoretical Foundations for Learning Science in Early Childhood

Chapter 1
Learning Science in Everyday Life – A Cultural-Historical Framework

Marilyn Fleer

Abstract Vygotsky's conception of the relations between everyday concepts and scientific concepts is introduced in this first chapter through an empirical example of a 6 year old child and an early childhood teacher making a slowmation animation together. The concept of force is examined. Through this example, cultural-historical theory is introduced as the theoretical perspective guiding the whole book. In this chapter, we do not simply introduce the chapters or sections of the book, but rather this chapter provides a theoretical framework and context where the major concepts and themes that are introduced later in the book are presented to give a comprehensive account of early childhood science education. The content of this book is introduced through a discussion of why this particular book on early childhood science is needed for advancing theory and practice in early childhood science education.

Keywords Cultural-historical theory • Early childhood science education • Constructivism • Slowmation • Children's science • Conceptual change • Force

1.1 Introduction: Our Historical Legacy

Sophia (6 years) was at park with her grandparents, two uncles and Chistina (teacher). She led Uncle JJ and Christine to a spinning piece of equipment that she discovered earlier. She sat on it and requested Uncle JJ to spin her.

Sophia:	*Spin!*
Christine:	*How should Uncle JJ do it?*
JJ:	*You should know how to spin it yourself…*
Sophia:	*Hold on tight*
JJ:	*Yes hold on tight (started to spin her) (Sophia grabbed on tightly to the pole)*
JJ:	*Later you'll be very giddy you know*
Sophie:	*Don't worry! (which also means: you can spin fast; I can handle it),(Uncle JJ turned the knob fast and Sophia squealed with excitement)*
JJ:	*Giddy already… Slow… The other way*
Sophia:	*Faster, faster, faster!*

M. Fleer, N. Pramling, *A Cultural-Historical Study of Children Learning Science*, Cultural Studies of Science Education 11, DOI 10.1007/978-94-017-9370-4_1

Christine:	*Is your head spinning too?*
Sophia:	*No, but the wheel is…*
Sophia:	*Faster and faster (squealed with excitement again)*
JJ:	*Slower…*

Sophia stood up and next was Uncle JJ's turn. He stood on the spinning equipment. Sophia held on to it with one hand and ran. She gave it a push and observed how it continued to spin without her holding on to it. He squatted down and the equipment slowed down. He exclaimed, "stop!" She helped by using her arm and body strength to bring him to a complete stop (Fleer & Hoban, 2013).

According to longstanding research in science education, very young children's everyday practical experiences of forces have been learned intuitively within specific everyday contexts, such as when Sophia is playing in the park and stops her uncle from continuing to spin (as shown in Fig. 1.1), or when she is spinning and comments (as discussed in the text above). These intuitive everyday practices have been linked closely to the explanations offered about why children do not hold the same views about concepts in science as scientists do. Over the past thirty years a huge volume of research has amassed in relation to children's thinking in science, most of which has drawn from constructivist views for informing learning and development. Even though there exists a large body of literature into understanding what children think, and what kinds of programs support children's learning in science, very little research attention has been

Fig. 1.1 Everyday understandings about *force* learned in the park

directed to children younger than five (Fleer & Robbins, 2003a; Robbins, 2009). This book seeks to redress this dearth in understanding – not by adding to what already exists, but rather by re-analysing this body of work and re-theorising understandings about our youngest scientific learner. To do this, we will draw specifically upon cultural-historical theory for our critique, followed by an examination of cultural-historical studies in science education (including our own recent research) to build our case (see Chap. 7). What is unique about our book is that we will use cultural-historical theory for better understanding the scientific conceptual development of very young children, going one step further than notable works, such as that of Eshach (2006) or Metz (2006). This chapter and the next, pave the way for this cultural-historical understanding of science learning by critiquing the existing literature and discussing those theoretical concepts thought important for the conceptual development of young children.

This chapter begins with a cultural-historical analysis of the longstanding science education research literature so that we better understand the research backdrop that has brought us to this point in time. We begin by an analysis of what is known about children's thinking in science, followed by a critique of the tools that have generated this body of knowledge over the past 30 years. All contemporary research contexts are laced with historically informed practices that contribute to how we currently enact and conceptualise early childhood science research. However, our analysis asks different kinds of questions about the nature of science learning than previous literature reviews because we are interested in determining the conceptual essence of very young children's learning and development in science as a collective endeavour. Our unit of analysis seeks to find the smallest cell that makes the biggest difference to young children's conceptual development (Davydov, 2008). We critique the concept of *children's science* because this work is still prevalent in early childhood science education, and its traces are still keenly felt. Later in this chapter we re-theorise this literature and the tools in relation to cultural-historical concepts.

1.2 Children's Science

Piaget's early work was instrumental in influencing a number of key researchers within the science education community by drawing attention to the explanations children give to natural phenomenon. As is well known, Inhelder and Piaget (1958) made particular kinds of naturalistic observations, and undertook specific conservation tasks, presenting the education community with particular types of interpretations. Some examples are presented, re-analysed and discussed in subsequent chapters.

What is particularly important for the focus of this book is how these ideas were taken up in the science education community in the late 1970s and early 1980s (see

Driver, Guesne, & Tiberghien, 1985; Osborne & Freyberg, 1985; White, 1988). Researchers at the time created specific tasks for children and older students to experience, with the view to ascertaining children's thinking in science for both primary (e.g., Roger Osborne in New Zealand) and secondary students (e.g., Ros Driver in the UK). Sadly, little research was done at the time in relation to children younger than 5 years of age. Examples of the kind of instruments that were designed to elicit children's thinking in science, included, concept maps, word associations, prediction-observation-explanation tasks, Venn diagrams, and a range of forms of interviews (White, 1988). For example, children were interviewed about an everyday incident that they would be familiar with (interview-about-incident) in order to determine their understandings about a concept, such as *force*. The children would be presented with a series of cards, such as the ones shown in Fig. 1.2. The children would be told that the interviewer was interested in their understanding about the concept – *force* – and that they would have a chat about the diagrams. The children

Fig. 1.2 Interview-about-incident (force) cards

would then be presented with a card and asked "Do you consider that there is a force on the ……………….., in your meaning of the word force?"; "Why do you say that?"; "Can you tell me more about that?" (Osborne, 1985, p. 43). Interviewers were initially puzzled by comments made, such as when a 9 year old was asked about if there is a force on the car, she replied: "No I don't think so because he is not forcing the car…the car won't move, it would be too heavy…he would be arguing at the car, kicking it…then he would probably start walking…he would do a force, if it was a brand new car…try and save his car instead of leaving it out" (p. 43). Noting the human aspect of the children's responses, as this 9-year-old has conceptualised, is central to understanding the comments children give in these contrived interview situations. Analyses at the time linked children's comments mostly to historical periods in which particular worldviews about the concepts were prevalent, for example, in physics, thinking from scientists, such as, Einstein, Newton and Buridan, who all had different theoretical explanations for the concept of *force*. For example:

Newtonian interpretation: (Is there a force on the bike?: "Yes…because something is trying to slow it down….because something is pushing it the other way so it slows down" (9 year old).

Buridanian interpretation: (Is there a force on the golf ball?: "The force from when he hit it is still in it" (13 year old).

Analyses of children's responses needed to broaden as more studies were undertaken, where findings noted how children associated *force* with an emotional state, like anger or feelings, and as 'something' that makes an object move, rather than causing a change in motion. Here, coercion, or physical movement, or muscular strength were common explanations given by young children – as mentioned in the spinning example of Sophia playing in the park. Frequently, children associated *force* with the object itself – as residing in the object (Buridanian interpretation), or as something that keeps an object moving. Importantly, the early research collectively showed that children generally only considered *force* in relation to the motion of an object and do not give thought to the *equal forces* that are acting to keep an object motionless (e.g., equilibrium situations).

Collectively, this research into children's thinking about *force* suggested that when children think about *force*, they consider it as the *property of an object*, rather than as the interaction between two or more objects. Children who think the former, generally give the following explanations:

- If something is moving, then there must be a *force* acting on it (or has acted on it)
- If an object is still, then there is no *force* acting on the object
- It is not possible to have movement if a *force* has not been applied
- If something is moving, then the *force* has determined its direction
- An object keeps moving because it has *force* inside of it to keep it going
- When something stops, then the *force* has been used up
- For older children, constant speed is associated with constant *force*

The outcomes of these early studies became known as *Children's Science*, and these early studies shaped the way science education research was conceptualized for almost 25 years. In addition, these early techniques used to elicit children thinking, such as the 'interview-about-incident' task, spawned a worldwide frenzy into finding out what children knew about a range of scientific concepts.

This research tells us that for 30 years, research within the science education community has revealed what very young children will express about their thinking for a range of scientific understandings, such as *force*, and this work shows that very young children will usually *explain science concepts at an intuitive level*. But a critique of how these techniques were used to elicit thinking, such as that undertaken by Margaret Donaldson (1978, 1992) when she re-examined Piaget's conservation tasks, is missing from the science education research literature (see also Richards & Light, 1986; Wood, 1988). Rather, the approaches used were deemed suitable for determining how children think about science concepts without questioning the reliability of the tools used for young children. We will return to this idea in Chap. 7 in this book when we look at how children are positioned in research.

The general science education research literature also shows that a range of concepts have been researched in the primary and secondary years, including light, electricity, heat and temperature, force and motion (this chapter), particulate nature of matter, chemical transformations, etc. In addition, more generic topics covering science concepts for younger children were also researched, such as when studying force (Chap. 1), light (Chaps. 2 and 10), geology (Chap. 8), animals and plants (Chaps. 6 and 8), magnetism, the human body (Chap. 11), floating and sinking, and night and day, plus more. We take up some of these science concepts and analyse these further later in this book.

Researchers initially published children's scientific thinking in international handbooks where understandings where discussed in relation to age levels and countries (i.e., see Duit, Treagust, & Widodo, 2008). In addition, most textbooks and scholarly books in science education bring these understandings about children's thinking together as a part of the source material that is deemed useful for informing the work of science teachers or primary and early childhood teachers working with science concepts (see early works such as, Driver et al., 1985; Osborne & Freyberg, 1985; and later publications, for example, Eshach, 2006; Harlen, 2010).

The science education research community has invested a great deal of effort and time into documenting these findings. Some thirty years of research have gone into finding out *children's science* or *alternative views* as they also became known as – due to signaling that children's scientific ideas were not wrong but rather were alternative to that which was accepted by the scientific community at that time (e.g., http://www.education.vic.gov.au/studentlearning/teachingresources/science/scicontinuum/monashdesign.htm). So what has underpinned this body of literature? What has been the basic logic and view of children's development that has guided researchers and educators in science education? Constructivism is the foundational theory that has guided us. Constructivism was informed by Piaget's own theory of child development (Vygotsky, 1987, p. 173). Constructivism is central to *Children's Science*. Children's science is:

- resistant to external suggestions
- deep within the child's thought
- age specific
- maintained in children's consciousness over several years
- the child's first answers

Over the past 10 years, we find that whilst there are still some papers published on children's science or children's alternative views, the quantity has significantly dropped. It is no longer the main area of interest to researchers. During this time, we have also seen more studies directed to the study of how very young children experience science education, including investigating their thinking in science as a result of participating in different kinds of programs, such as *Interactive teaching* (e.g., Kirkwood, Bearlin, & Hardy, 1989), *developmentally appropriate science* and *constructivist approaches* (e.g., Martin, Jean-Sigur, & Schmidt, 2005), a *process approach* (Kirch, 2007), and *cultural constructions* of science and *children's interests* as basis for science instruction (e.g., Siry & Kremer, 2011; Siry & Lang, 2010). We also see fusion between teaching philosophies and science, such as use of Reggio Emilia (Stegelin, 2003), emergent curriculum with science standards (Baldwin, Adams, & Kelly, 2009), Developmentally Appropriate Programs (DAP) and pedagogical practices in science (Hadzigeorgiou, 2001). These ideas will be taken up further in the book.

A careful analysis of the research and the central theoretical frameworks that have progressed the science education field for the past 30 years reveal *a rather one-sided approach to examining children's scientific thinking*. To step back from this plethora of findings, and to think about children's science in a different way, requires another way of considering the traditional view of the process of mental development of children. We need to move beyond *age as a central criterion* for scientific progression (e.g., http://www.deakin.edu.au/arts-ed/education/sci-enviro-ed/early-years/floating.php) or *historical periods in science* for explaining particular types of thinking (see Osborne, 1985), but rather we need to think in new ways about what we already know. Vygotsky (1997) argued that "…it is easier to assimilate a thousand new facts in any field than *to assimilate a new point of view of a few already known facts*" (p. 1; our emphasis). Assimilating another fact about how children think in science does not progress the field, because it would seem that much of the published research has resulted in little progress being made in dealing with the outcomes of these views children hold (Skamp, 1993). That is, what do we do with this expansive knowledge base about children's science or the record of alternative views? How can this help us with understanding the conceptual development of very young children? We need to think about children's science in a different way. We need to move from children's thinking in science, to a cultural-historical reading of the *thinking and learning child,* where we go beyond the lone child's construction of knowledge, and consider the relations between the child and their social and material environment, where the teacher plays an important role.

A cultural-historical reading of science education would position science as a form of cultural knowledge that is historically and collectively formed and understood, rather than as something that is located within the individual. Research into the concepts that children hold, particularly those studies that are framed from a constructivist tradition, would invite the researcher to look closely at what it is an individual child says and believes at a particular moment or point in time. That is, it is an individually held construction of science concepts/knowledge. Here scientific knowledge is broken down into individual concepts, decomposing the whole into parts, as the basis for investigating what a child thinks about an individual concept. Yet we know that children traditionally do not think or learn in reductionist ways. In many respects, this kind of science education research can best be summed up by drawing upon the words of Vygotsky (1997) to say that constructivist research into young children's concepts in science is like a "mosaic of mental life developed comprised of separate pieces of experience, a grandiose atomistic picture of the dismembered human mind" (P. 4). When in fact science is a system of concepts, invented by humans, and which are given meaning through the Western paradigm of science (see Aikenhead, 2006; Aikenhead & Michell, 2011) and enacted with others. These ideas are discussed further in Chap. 6. Knowing about what children think in science is only one side of the '*teaching-learning* coin' for the development of scientific concepts. A cultural-historical reading of conceptual development would suggest that knowledge is not contained within the individual head, but rather it is distributed across people, who collectively contribute to thinking in science, whether this is at a large scale policy level, such as, a Government initiative (e.g., science to inform sustainability by reducing carbon emissions through painting the roof of all CBD buildings white to reduce energy consumption for cooling), or whether it is about how to change family practices in a home for keeping everyone warm in winter (i.e., insulate the walls or close doors and use draft stoppers on the base of doors), or how a child stays warm on their way to preschool by putting on a coat, gloves and hat. Reducing scientific concept formation to the individual as an explanation of 'what they know' or 'think in science' also creates a binary that dates back to Descartes 'mind – body' split. This form of Cartesian logic has been extensively critiqued and found wanting (see Chap. 5 for discussion of forms of knowledge and Chap. 11 on representation in science).

Rogoff (1995, 1998, 2003) has also discussed the *artificial boundary* between the internal mind and external world that is formed when Cartesian logic is used to explain how someone thinks and learns. The binary of externalization and internalization that results is explained by traditional psychology:

> Developmental research has commonly limited attention to either the individual or the environment – for example, examining how adults teach children or how children construct reality, with an emphasis on either separate individuals or independent environmental elements as the basic unit of analysis (Rogoff, 1995, p. 139).

A cultural-historical reading would not separate out individuals, but rather would focus on the interdependence between them, where the unit of analysis would preserve the essence of their social and material interactions, rather than separating out how an individual child thinks about one concept or another. The child, the interactions between others, and the cultural and historical context in which the children/adults are operating does not exist separately. An analogy with the concept of *force* can be made. It would be a nonsense to examine only one source of a *force*, disembedded from the system of forces that were operating, in order to understand the concept of *force*. *Force* can only be understood within a system of *forces* that are acting, even when the object under investigation is stationary. In a cultural-historical reading of the science education literature, we must consider the mutually constituting processes that are at play when we consider conceptual development of children, and not the individual understandings of scientific concepts. It is incomplete to only focus on an individual's conceptual development, without also considering the social interactions between individuals in coming to speak about or to discuss a particular science concept. A child does not develop in isolation. Importantly, it is also not complete to examine the cultural community within which the conceptual development is being foregrounded, because communities also develop, they are not static. Logically, children's science puts children's thinking in opposition with adult thinking in science, and this is unhelpful as it means in teaching, only the 'replacement of children's thinking to that of the adult' must occur. Rogoff (1995) has used the concepts of *guided participation* and *participatory appropriation*, and the metaphor of *apprenticeship* to explain how development occurs in order to move beyond a model of externalization and internalization of concepts, and this is helpful for thinking outside of the binary that Children's Science has inadvertently created. She shows the binary graphically through the portrayal of the boundary that is formed when first, a behaviourist view is considered, and second, a constructivst view of teaching and learning is featured (see Fig. 1.3).

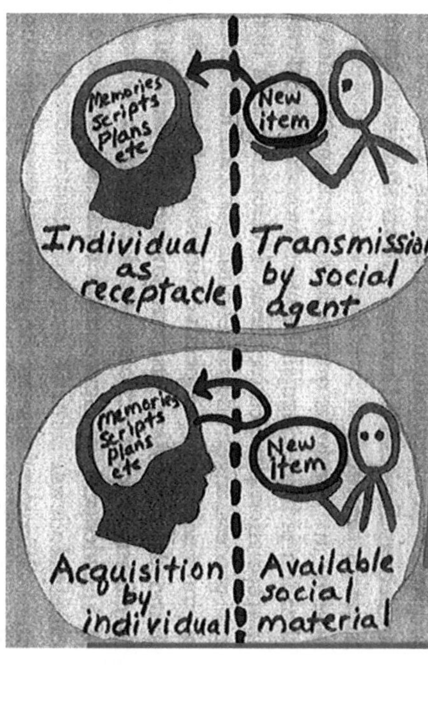

Fig. 1.3 Boundary between internal and external (Extract from Rogoff, 1995, p. 154)

Further to this, we can see in this representation, but also in the critique given by Vygotsky (1987) on Piaget's spontaneous and non-spontaneous concepts which informed constructivism, that children's expression of spontaneous concepts (alternative views) is distinctly different to the accepted scientific concepts. The focus in this line of enquiry is on the *differences* rather than the *connection*. Piaget views "the development of concepts as a mechanical combination of two separate processes, processes which have nothing in common and move, as it were, along two completely isolated or separate channels" (Vygotsky, p. 174). *Children's science* is viewed as completely different to, and disconnected from, the accepted scientific view.

Locating science knowledge within the individual mind is a limited view of science and therefore science learning. Yet, this perspective underpins much of the research that has flowed from *Children's Science* or children's *alternative views*, and into the models of teaching which focus on *conceptual change*. See Chap. 7 for details of how children are positioned in research and how individual understandings of scientific ideas are perpetuated. We now turn to the other side of the coin, and focus on the research into the teaching of science.

1.3 Conceptual Change and Socioscientific Approaches to Teaching Early Childhood Science

The other side of the coin which has steadily gained momentum within the science education literature is socioscientific approaches to teaching. This perspective sees children engaged in a variety of social dilemmas that they must solve. These dilemmas are usually conceptual, procedural or technological. Real world contexts create the conditions for engaging students in science learning as children discuss and debate the products or processes of science. Examples are usually controversial, and often include cloning, climate change, genetically modified food, soil degradation, and stem cell research. From this literature we notice a growing number of studies which focus on how children learn science when a socioscientific perspective is featured. According to Sadler and Zeidler (2005) "the socioscientific issue movement arises form a conceptual framework that unifies the development of moral and epistemological orients of students and considers the role of emotions and character as key components of science education" (p. 113). Through engagement in controversial issues, it is argued that children develop scientific literacy. Scientific literacy entails children being sufficiently informed to be able to make decisions about societal needs and practices. A focus on scientific reasoning and conceptual change of children is central. Examples include enhancing the quality of argumentation in school science (Osborne, Erduran, & Simon, 2004), and argumentation in science (Naylor, Keogh, & Downing, 2007).

This perspective has been driven by finding out what engages children (and adults) in science learning (or teaching this subject) or how effective is the socioscientific perspective for framing learning and teaching in order to *change children's thinking*. Here the focus is still on *conceptual change* as the ultimate marker of program success. However, this perspective is still located within a constructivst view, one this is conceptually coherent with *Children's Science.* It is not a change in theory or practice, but rather it is an artifact of the original work done in 1970s and 1980s. We take up the variety of examples in the third section of the book where we discuss these ideas in more detail.

Research into socioscientific models of teaching at all education levels (except early childhood) highlights the affective dimensions of morality in scientific reasoning (Sadler 2004a, 2004b; Sadler & Zeidler, 2005) and the cognitive conflicts that arise, as well as both raising the importance of context (Korpan, Bisanz, Bisanz, & Henderson, 1997; Sadler & Zeidler, 2005) and student engagement in science (Bell & Linn, 2000; Sadler & Zeidler, 2004). In the extensive research of Sadler and Zeidler (2005) into informal scientific reasoning they have noted the importance of a personal dimensions of decision making in socioscientific dilemmas of college students. In particular they found that personal experiences, emotive engagement and the social and moral issues associated with informal reasoning as drivers in scientific thinking and decision making. Knowledge of the social world rather than material world was central to student reasoning in science. Morality has been thought to lead this thinking, as noted by Bell and Lederman (2003). However, Sadler and Zeidler found that attempts at isolating "morality, and by extension personal or social factors, as a guiding factor in the determination of positions regarding socioscientific issues are misguided" (p. 129). They found that reasoning in science was subsumed by morality, personal experiences, emotive factors and social considerations. This finding is highly significant because it draws attention to the unity of emotions and cognition in scientific learning, something that is foregrounded by Vygotsky in his earlier writings (Vygotsky, 1991) and the final works (Vygotsky, 1987). We discuss the unity of affect and intellect later in the book (see Chap. 3).

As might be expected, a socioscientific approach to teaching science has not directly influenced research in early childhood education. Presenting scientific dilemmas to children in the birth to five year period has really only emerged in relation to environmental sustainability (e.g., Davies, Engdahl, Otieno, Pramling-Samuelson, & Siraj-Blatchford, 2009) where research is focused mostly on what children think and how they might act, rather than investigating how through argumentation young children resolve issues or take a personal stand in relation to a socioscientific dilemma. Further to this research into socioscientific approaches to teaching science, are those studies which have examined everyday situations and contexts as either motivating learners, or influencing their concepts in science. However, most of these studies focus on secondary students and are outside of the scope of this book.

Collectively, these studies represent the other side of the coin – the teaching context. What these studies miss is the *dialectical relations* between context and concepts. That is, studies of children's thinking is only one source of knowledge and understanding of science education, and this does not go far enough in understanding how scientific knowledge is constructed by children. Research into teacher knowledge or program effectiveness also does not go far enough, as the focus of research attention is generally in one direction - from the program *to the child*. These programs focus on how their strategies or intentions shape children's scientific understandings. They generally do not examine how children shape the programs themselves because the focus is 'on the program'. This is the same for studies into teacher education, where the research either examines pre-service teachers scientific knowledge or competence (e.g., Garbett, 2003, 2007), or examines what effect a tertiary program has had on teacher knowledge or competence to teach science (Appleton, 1995; Gilbert, 2009). The studies do not examine how pre-service teachers shape the programs they experience or how teacher courses take account of the student as the learner of science. Simply studying the effect of a science program on children's learning is like only studying an embryo. The essence of development is there, but there is much to still know about how the embryo develops over the course of its lifespan. An investigation of the embryonic development of science programs is a limited approach to the study of science education. We believe that in the field of research into early childhood science education we are still at the embryonic level of development.

Cultural-historical research into science education seeks to examine the relations between the child/teacher and the concepts/contexts as a dialectical process, where the learner is shaped by, but also shapes the social and material conditions for science learning. We now turn to a discussion of the central theoretical concepts for understanding this dialectical relation.

1.4 The Dialectical Relations Between Everyday Concepts and Scientific Concepts

The experiences of Sophia and Uncle JJ in the park, led Sophia and her teacher Christine into preparing a slowmation animation together in order to explain the *forces* acting in the park. Hoban (2007) explains that *Slowmation* (www.slowmation.com) is a simplified way of making stop-motion animation. By taking a series of photographs of objects and loading them into Slowmation software, the images can be played slowly at 2 fps allowing Sophia and her teacher to narrate an explanation of a science concept of *force* over their 'set of slow moving images' of their recreation of the park scene using two soft toys (Fig. 1.4). Their narration follows:

Sophia:	*Yay, see-saw, I can go up and down. Why can't I go up?*
Christine:	*Hello Doggy, that's because gravity pulls you down. But I'm heavier than you. See, now I go down and you go up. Just give a little kick and we will go up and down, up and down.*
Sophia:	*Thank you, but now I want to play on the slide. I am climbing up. Wheee. See, I came down so fast. Ah-ha, that's because gravity pulls me down.*

Fig. 1.4 Slowmation animation co-constructed by Christine and Sophia

Christine:	*You are right, Doggy.*
Sophia:	*Watch me, OUCH…*
Christine:	*Oh dear, be careful, Doggy.*
Sophia:	*Okay. Shall we play on the merry-go-round, Piggy?*
Christine:	*Sure, why not? Do you wanna go fast or slow?*
Sophia:	*Fast, very very fast!*
Christine:	*Okay, then I will have to give you a very big push! Watch out!*
Sophia:	*Oh! I am spinning so fast! I can't stop! HELP!*
Christine:	*Okay, let me give you a little push in the opposite direction then.*
Sophia:	*Oh man, I am so giddy.*
Christine:	*Ha-ha, I'm sure. Hey look, there's a football, shall we play? I'll kick it to you. Kick it back to me.*
Sophia:	*Ha (kicking action).*
Christine:	*Doggy, you kicked it too lightly. Try again.*
Sophia:	*Okay.*
Christine:	*Oh-oh, you have to kick even harder than this.*
Sophia:	*Ha (kicking action).*
Christine:	*Yay, that's good. Now back to you! Now I am going further away. See if you can make the ball travel far and fast.*
Sophia:	*No problem, I will kick it really hard. There you go. Piggy, you went the wrong way.*
Christine:	*He-he, oopsy, let me try again.*
Sophia & Christine:	*Aren't we having so much fun?*
Sophia:	*Gravity pulls you down.*
Christine:	*Gravity pulls you down.*
Sophia:	*A hard push will make something spin fast.*

Christine:	*A light push will make something spin slow.*
Sophia:	*A light kick will make the ball move a little.*
Christine:	*A strong kick will make the ball travel far.*
Sophia & Christine:	*The End. Hope you enjoyed the show* (Adapted from Fleer & Hoban, 2013, p. 66).

We deliberately introduce this Slowmation example as an explicit investigation of the scientific concept of *force* by a young child and her teacher because we use it to introduce a different conceptualization of early childhood science education. At the beginning of this chapter we shared the tacit intuitive engagement with the concept of *force* that is experienced by Sophia in the park, where Uncle JJ begins by spinning Sophia. Uncle JJ says, "You should know how to spin it yourself…" Sophia responds by saying, "Hold on tight" as Uncle JJ beings to spin her whilst she says, *"Yes hold on tight".* Sophia holds tightly onto the pole.

Rather than conceptualizing this reading of Sophia's experiences as part of building an intuitive understanding of *force* as explained by the concept of *Children's Science,* or potentially building an *alternative view of force,* we prefer to conceptualise this as Sophia's *everyday concept* of *force*. Here we use Vygotsky (1987) concept of everyday concepts and scientific concepts for our analysis. Vygotsky argued that "*The existence of a concept does not coincide with consciousness of that concept either in the moment of its appearance or in its mode of functioning.* The former may appear earlier than the latter and act independently of it. Analysis of reality on the basis of the concept emerges much earlier than analysis of the concept itself" (p. 161; Original emphasis). Much of the alternative conceptions literature suggested that children's ideas in science get in the way of learning scientific concepts, and once formed they are difficult to 'remove' (see critique by Fleer & Robbins, 2003b) as discussed previously. However, a cultural-historical reading would suggest that it is important for children to develop everyday concepts of their experiences, as these may arise earlier than scientific concepts, or later than scientific concepts, or even at the same time. But regardless of when an everyday concept is formed, everyday concepts **are central**, *not alternative*, for developing a scientific concept. For instance, a cultural-historical analysis seeks to capture both the everyday concept of *force*, alongside of how the child gains a conscious awareness of the concept. We begin to see this consciousness emerging in the everyday situation of Sophia spinning, and Christine using everyday language to draw her attention to the experience *"Is your head spinning too?"* Sophia does not respond to this, but rather references the concrete situation by saying, *"No, but the wheel is…".* Vygotsky has argued that "We know from research on concept formation that the concept is not simply a collection of associative connections learned with the aid of memory. We know that the concept is not an automatic mental habit, but a *complex and true act of thinking* that cannot be mastered through simple memorization. The child's thought must be raised to a higher level for the concept to arise in consciousness.

Table 1.1 Everyday concepts and scientific concepts

<table>
<tr><th>Everyday concepts</th><th>Scientific concepts</th></tr>
<tr><td>Sophia knows that Uncle JJ can make her spin faster; she uses the term ‘faster’</td><td rowspan="3">Force is required to get a still object to start moving. The greater the force, the greater the speed. However, the friction between the object and the surface slows down the speed. Force exerted in opposite direction of the moving object will cause it to slow down or stop.</td></tr>
<tr><td>Sophia is aware that when it spins too fast, there is a danger of falling off</td></tr>
<tr><td>Sophia knows that she can stop Uncle JJ from moving by using her arms and body strength</td></tr>
</table>

Adapted from Fleer and Hoban (2012, p. 65)

At any stage of this development, the concept is *an act of generalization*” (p. 169; Original emphasis). In this reading of children’s thinking, everyday concepts build in breadth, depth and complexity as the child experiences more. In Table 1.1 we see the generalization of the everyday concept of force that is being lived as Sophia plays in the park with her Uncle JJ. But we also know that there is a scientific explanation of force as detailed on the right, as part of Christine’s teaching program of intentional teaching of the concept of force to Sophia (as planned by using Slowmation). Vygotsky argued that “The process of defining the concept when it is torn from the concrete situation in which it was developed, when it no longer depends on concrete impressions and begins to develop in an entirely abstract plane, is significantly more difficult” (p. 161). Scientific concepts do not by their very nature have built into them some internal form of logic that can be reproduced by the child (ie as though somehow moving from Buridan to Newtonian explanations of force), but rather they are cultural inventions that are inherited by children from the society in which they grow up. We capture scientific concepts through truncated and specialized terms, formulae and explanations. They are built in practice, named and then abstractly used away from their original site of conceptualisation. Therefore the limitation of everyday concepts must lie in ‘its *incapacity for abstraction’* because without naming the practice and giving it an explanation, it cannot be transported to another context. Conversely, when the scientific term is introduced to a child away from its site of use or development, then the weakness of the scientific concept lies in its “*verbalism*, in its insufficient saturation with the concrete” (p. 169).

Vygotsky suggested that everyday concepts lay the foundations for scientific thinking. However, he also said that scientific concepts lay the foundations for everyday conceptual thinking. The question is how are these two concepts interrelated so that they become foundations for each other? Vygotsky used the concept of a shadow to eloquently present this challenge to us: “What is the relationship between instruction, learning, and the processes involved in the internal development of scientific concepts in the child’s consciousness? Are these simply two aspects of what is essentially one and the same process? Does the process involved in the internal development of concepts follow instruction like a shadow follows the object which casts it, not coinciding with it but reproducing and repeating its movement, or do both processes exist in a more complex and subtle relationship which requires special investigation?" (p. 169).

Vygotsky (1987) has argued that the *direct* teaching of a concept is 'pedagogically fruitless'. When teaching of science occurs through verbal instruction or by giving scientific explanations of particular concepts, such as *force*, away from the everyday context, then this "achieves nothing but mindless learning of words" and under "these conditions the child learns not the concept but the word and this word is taken over by the child through memory rather than thought" (p. 170). Conversely, the experiences of the concept in everyday life, whilst providing a valuable intuitive and sensory based experience, does not on its own lead to conceptual development or the understanding of a concept such as *force*. Rather, it is the relations between everyday concepts and scientific concepts that leads to conceptual development. According to Vygotsky:

> These two types of concepts are not encapsulated or isolated in the child's consciousness. They are not separated from one another by an impenetrable wall nor do they flow in two isolated channels. They interact continually. This will inevitably lead to a situation where generalizations with a comparatively complex structure – such as scientific concepts – elicit changes in the structure of spontaneous concepts. Whether we refer to the development of spontaneous concepts or scientific ones, we are dealing with the development of a unified process of concept formation. By its very nature, however, it remains a unified process. It is not a function of struggle, conflict, or antagonism between two mutually exclusive forms of thinking. (p. 177).

In contrast, conflict between concepts underpins not only the work of Piaget (i.e. cognitive conflict), but has been used for many socioscientific teaching programs which seek to foster argumentation and create debate around general societal issues. This is the case mostly for programs designed for students outside of the early childhood period. In this theoretical reading "it is only the *child's spontaneous concepts and representations* which can serve as the source of direct knowledge of the unique qualities of the child's thought" (Vygotsky, 1987, p. 174; Our emphasis), and these are generally considered in relation to the views of the adults which surround the child. In separating out everyday concepts from scientific concepts, this implies that they have no relationship to each other (as the boundary in Fig. 1.2 shows). Here too, the practical value of the everyday concepts held by the child is not considered, other than as something in need of conceptual change.

A cultural-historical reading of the relations between everyday concepts and scientific concepts places great value on the everyday concepts, because it is through the act of capturing the everyday experiences, naming these through word labels in practice, that the child begins to develop a conscious realization and potential generalization for these everyday concepts. "When the child first learns a new word the development of its meaning is not completed but has only begun". (Vygotsky, 1987, p. 170). Vygotsky has shown that "the word acts as a means of forming the concept" (p. 165), and this is only part of the process of conceptual development.

According to Vygotsky scientific concepts can arise only "on the foundation provided by the lower and more elementary forms of generalization which previously exists. They cannot be simply introduced into the child's consciousness from the outside" (p. 177). Here we see the significance of the relations between everyday concepts to scientific concepts. However, what is the relations between the scientific concepts to everyday concepts? What does this means for science education? This relation can best be captured through an analogy, first cited by Vygotsky in relation to learning a new language. A child speaks her/his mother tongue, but has no conscious awareness of the grammatical structures and forms of speech that are used. It is only through the study of a second language, that the child's attention is drawn to the forms of speech and the grammatical structures inherent in the new language. It is through explicitly examining the second language that it is possible to recognize the verbs, nouns, adjectives, and grammatical structures of the mother tongue. But it is precisely because the mother tongue is already well developed that it becomes possible to recognise its basic structure and a greater level of awareness or consciousness arises for the child, previously not possible. Although the pathways to learning are different, their relations are tightly fused. One gives meaning to the other. The relations between everyday concepts and scientific concepts operate in exactly the same way. Introducing scientific concepts support the awareness raising of concepts that are enacted in everyday practice. New meaning is created in everyday practice as scientific concepts are introduced and conceptual development occurs. However, without the everyday concepts in practice, no meaning can be given to the words of the scientific concepts. Scientific concepts are given meaning through these everyday experiences. Conscious awareness of scientific concepts is clearly a special process, but what does this really mean in science based practice?

If we examine the slowmation example at the beginning of this section we can see that Christine sought to explicitly bring together Sophia's everyday concepts with scientific concepts. She did this by reproducing the play experience as a slowmation animation. What is important here, is that Christina and Sophia had to think about the everyday experiences of playing in the park and to create a representation of these in some way. The soft toys and the plasticine provided the resources for animating the experiences. This level of abstraction is still located within the everyday experiences of the child, as connected to the concrete lived event. However, the creation of a script for narration, not only provided the means for making aspects of the experience more conscious, it provided the mechanism for introducing a scientific explanation of the everyday concepts. For example, in the scene of the see-saw, Sophia says in the narration, `"Yay, see-saw, I can go up and down. Why can't I go up?"`, and Christine responds `"Hello Doggy, that's because gravity pulls you down. But I'm heavier than you. See, now I go down and you go up. Just give a little kick and we will go up and down, up and down."`

Labeling the different sources of *force* in a park as occurs in the narration, and through the actual signs that are used to punctuate the significance of the forces that are acting, this creates a dialogue that moves the child's thinking from the everyday concepts and to the scientific concept of *force*. Having all of the sources of *force* that are acting during a particular everyday experience, such as the see-saw, is as important, as undertaking an audit of the forces that are acting as a result of all of the equipment, particularly equipment that is not moving. The latter being the more difficult idea for young children to contemplate. The scientific concept is more than a word. The development of a scientific concept captures the complexity of the movement between the everyday concepts and the scientific concept, and together over time the young child begins to think and act using scientific concepts.

Here the relations between learning and development are important. In this book we draw upon Hedegaard and Fleer's (2013, p. 183) conceptualisation of learning as a change in the child's "relation to another person and activities in specific settings" as a result of learning science concepts. The child is able to think and act differently in their world. We define development as a process where "children's motive orientation and engagement in different activity settings change qualitatively" and as such their leading motive changes (Hedegaard and Fleer, 2013, p. 183). In early childhood this usually means a change from a play motive to a learning motive. But these constructs of learning and development are interrelated. We examine the relations between learning and development throughout this book through examples taken from our research and the research of others. In the final chapter we discuss this relation explicitly in the context of introducing a model for science teaching suitable for early childhood settings.

In the next chapter we turn our attention to the child as a learner within the preschool environment, in order to determine how the resources and the structures support learning in science.

References

Aikenhead, G. S. (2006). *Science education for everyday life: Evidence-based practice*. New York: Teachers College Press.

Aikenhead, G., & Michell, H. (2011). *Bridging cultures: Indigenous and scientific ways of knowing nature*. Toronto, Canada: Pearson Canada.

Appleton, K. (1995). Student teachers' confidence to teach science: Is more science knowledge necessary to improve self-confidence? *International Journal of Science Education, 17*(3), 357–369.

Baldwin, J. L., Adams, S. M., & Kelly, M. K. (2009). Science at the centre: An emergent, standards-based child-centered framework for early learners. *Early Childhood Education Journal, 37*, 71–77.

Bell, R. L., & Lederman, N. G. (2003). Understandings of the nature of science and decision making on science and technology based issues. *Science Education, 87*(3), 352–377.

Bell, P., & Linn, M. C. (2000). Scientific arguments as learning artifacts: Designing for learning from the web with KIE. *International Journal of Science Education, 22*, 797–817.

Davies, J., Engdahl, I., Otieno, L., Pramling-Samuelson, I., & Siraj-Blatchford, J. (2009). Early childhood education for sustainability: Recommendations for develop. *International Journal of Early Childhood, 41*(2), 113–117.

Davydov, V. V. (2008). *Problems of Developmental Instruction: A theoretical and experimental psychological study. International perspectives in non-classical psychology*. New York: Nova.

Donaldson, M. (1978). *Children's minds*. Glasgow, Scotland: Fontana/Collins.

Donaldson, M. (1992). *Human minds: An exploration*. London: Penguin.

Driver, R., Guesne, E., & Tiberghien, A. (1985). *Children's ideas in science*. Buckingham, UK: Open University Press.

Duit, R., Treagust, D., & Widodo, A. (2008). Teaching science for conceptual change: Theory and practice. In S. Vosniadou (Ed.), *International handbook of research on conceptual change* (pp. 629–646). New York/London: Routledge.

Eshach, H. (2006). *Science literacy in primary schools and pre-schools*. Dordrecht, The Netherlands: Springer.

Fleer, M., & Hoban, G. (2013). Using "Slowmation" for intentional teaching in early childhood centres: Possibilities and imaginings. *Australasian Journal of Early Childhood, 37*(3), 137–146.

Fleer, M., & Robbins, J. (2003a). Understanding our youngest scientific and technological thinkers: International developments in early childhood science education, Editorial. *Research in Science Education, 33*, 399–404.

Fleer, M., & Robbins, J. (2003b). "Hit and run research" with "hit and miss" results in early childhood science education. *Research in Science Education, 33*, 405–432.

Garbett, D. (2007). Assignments as a pedagogical tool in learning to teach science: A case study. *Journal of Early Childhood Teacher Education, 28*, 381–392.

Garbett, D. (2003). Science education in early childhood teacher education: Putting forward a case to enhance student teachers' confidence and competence. *Research in Science Education, 33*, 467–481.

Gilbert, A. (2009). Utilizing science philosophy statements to facilitate K-3 teacher candidates' development of inquiry-based science practice. *Early Childhood Education Journal, 36*, 431–438.

Hadzigeorgiou, Y. (2001). The role of wonder and 'romance' in early childhood science education. *International Journal of Early Years Education, 9*(10), 63–69.

Harlen, W. (Ed.). (2010). *Principles and big ideas of science education*. Herts, UK: Association for Science Education.

Hedegaard, M., & Fleer, M. (2013). *Leaning, play and children's development: How families and schools shape children's learning*. New York: Cambridge University Press.

Hoban, G. (2007). Using Slowmation for engaging preservice elementary teachers in understanding science content knowledge. *Contemporary Issues in Technology and Teacher Education, 7*(2), 1–9.

Inhelder, B., & Piaget, J. (1958). *The grown of logical thinking from childhood to adolescence*. London: Routledge & Kegan Paul.

Kirch, S. A. (2007). Re/Production of science process skills and a scientific ethos in an early childhood classroom. *Cultural Studies of Science Education, 2*, 785–845.

Kirkwood, V., Bearlin, M., & Hardy, T. (1989). New approaches to the inservice education in science and technology of primary and early childhood teachers (or mum is not dumb after all!). *Research in Science Education, 19*, 174–186.

Korpan, C. A., Bisanz, G. L., Bisanz, J., & Henderson, J. M. (1997). Assessing literacy in science: Evaluation of scientific news briefs. *Science Education, 81*, 515–532.

Martin, D. J., Jean-Sigur, R., & Schmidt, E. (2005). Process-oriented inquiry-a constructivist approach to early childhood science education: Teaching teachers to do science. *Journal of Elementary Science Education, 17*(2), 13–26.

Metz, K. (2006). Rethinking what is "developmentally appropriate" from a learning progression perspective: The power and the challenge. *Review of Science, Mathematics and ICT Education, 3*(1), 5–22.

Naylor, S., Keogh, B., & Downing, B. (2007). Argumentation and primary science. *Research in Science Education, 37*(1), 17–39.

Osborne, J., Erduran, S., & Simon, S. (2004). Enhancing the quality of argumentation in school science. *Journal of Research in Science Teaching, 41*(10), 994–1020.

Osborne, R. (1985). Building on children's intuitive ideas. In R. Osborne & P. Freyberg (Eds.), *Learning in science: The implications of children's science* (pp. 41–50). Auckland, New Zealand: Heinemann.

Osborne, R., & Freyberg, P. (1985). *Learning in science: The implications of children's science*. Auckland, New Zealand: Heinemann.
Richards, M., & Light, P. (Eds.). (1986). *Children of social worlds*. Oxford, UK: Blackwell.
Robbins, J. (2009). Analysing young children's thinking about natural phenomena: A sociocultural/cultural historical perspective. *Review of Science, Mathematics and ICT Education, 3*(1), 75–97.
Rogoff, B. (1998). Cognition as a collaborative process. In D. Kuhn & R. S. Siegler (Eds.), *Cognition, perception and language* (Vol. 2, Handbook of child psychology, W. Damon (Ed.), 5th edn) (pp. 679–744). New York: Wiley.
Rogoff, B. (1995). Observing sociocultural activity on three planes: Participatory appropriation, guided participation and apprenticeship. In J. V. Wertsch, P. Del Rio, & A. Alvarez (Eds.), *Sociocultural studies of mind* (pp. 139–164). Cambridge, MA: Cambridge University Press.
Rogoff, B. (2003). *The cultural nature of human development*. Oxford, UK: Oxford University Press.
Sadler, T. D. (2004a). Moral and ethical dimensions of socioscientific decision-making as integral components of scientific literacy. *The Science Education, 13*, 39–48.
Sadler, T. D. (2004b). Informal reasoning regarding socioscientific issues: A critical review of the literature. *Journal of Research in Science Teaching, 4*, 513–436.
Sadler, T. D., & Zeidler, D. L. (2004). The morality of socioscientific issues: Construal and resolution of genetic engineering dilemmas. *Science Education, 88*, 4–27.
Sadler, T. D., & Zeidler, D. L. (2005). Patterns of informal reasoning in the context of socioscientific decision making. *Journal of Research in Science Teaching, 42*(1), 112–138.
Siry, C. A., & Lang, D. E. (2010). Creating participatory discourse for teaching and research in early childhood science. *Journal of Science Teacher Education, 21*, 149–160.
Siry, C., & Kremer, I. (2011). Children explain the rainbow: Using young children's ideas to guide science curricula. *International Journal of Science Education and Technology, 20*, 643–655.
Skamp, K. R. (1993). Research themes, styles, purposes and future directions. In D. Goodrum (Ed.), *Science in the early years of schooling: An Australasian perspective* (pp. 43–63). Perth, Western Australia: Curtin University of Technology.
Stegelin, D. A. (2003). Application of the Reggio Emilia approach to early childhood science curriculum. *Early Childhood Education Journal, 30*(3), 163–169.
Vygotsky, L. S. (1987). Problems of general psychology. In R. W. Rieber, & A. S. Carton (Eds.), *The collected works of L.S. Vygotsky* (N. Minick, Trans., Vol. 1). New York: Plenum Press.
Vygotsky, L. S. (1991). *The psychology of art*. Cambridge, MA: MIT Press.
Vygotsky, L. S. (1997). *The collected works of L. S. Vygotsky, Volume 4: The history of the development of higher mental functions* (M. J. Hall, Trans., R. W. Rieber, Ed.). New York: Plenum Press.
White, R. T. (1988). *Learning science*. Oxford, UK: Basil Blackwell.
Wood, D. (1988). *How children think and learn*. Oxford, UK: Basil Blackwell.

Chapter 2
How Preschools Environments Afford Science Learning

Marilyn Fleer

Abstract This chapter specifically examines how science learning is afforded as a result of children being in preschool environments. Wondering is introduced as a way of conceptualizing how young children and teachers can interact to support science learning. The literature is examined in order to determine what science learning is possible through the different areas within the outdoor area, and the different areas within the centre. The concept of sciencing is drawn out of the literature and is used for analysis in a study of 3–5 year old children's learning science. Formal sciencing [composting (decomposition]), informal sciencing (prism on window sill [refracting light]) and incidental sciencing (textured path and chalk [force]), are discussed. The research introduced also noted how science can be foregrounded as part of the traditional areas within the preschool (Sensory garden [herbs – use, growth and care]). In addition, it was noted that science areas can be specifically organised through building science infrastructure into the centre (light area [blocking light, light reflecting and refracting]). Importantly, this chapter also shows how the using of science in everyday life in the centre (e.g. weather watching) affording science learning amongst preschool children.

Keywords Everyday concepts and scientific concepts • Light • Sciencing • Wondering

2.1 Introduction

In most play-based settings, teachers draw upon the centre environment (indoors and outdoors) for supporting children's learning. In these settings there are already many opportunities for science exploration to occur, without the need for specifically planned and organized science experience. How children experience this environment is central for determining if and how children are oriented to science learning.

In keeping with the theoretical focus of this book, we examine this central question of experience from a cultural-historical perspective. To do this we begin by drawing upon the theoretical concepts from a lecture given by Vygotsky, and originally published in 1935, that specifically examines 'the problem of the environment' (Vygotsky, 1994). Environment refers to both the material and social context

M. Fleer, N. Pramling, *A Cultural-Historical Study of Children Learning Science*, Cultural Studies of Science Education 11, DOI 10.1007/978-94-017-9370-4_2

of the child, where affective relations to science are featured. This is followed by a broader conceptual discussion, which uses more contemporary sources, and empirical material to illustrate how preschools afford science learning.

2.2 Cultural-Historical Development of Science

In the previous chapter it has been argued that scientific knowledge is a human construction which is passed from one generation to the next, where humans transform, reject, evolve and apply these knowledge systems to everyday life and research. For instance, when a mother says to an 8-year-old child that the colours in a rainbow that they can see are the result of light being refracting through a series of rain droplets, then she is passing on scientific knowledge acquired historically. The mother has not personally invented this explanation, but has herself acquired this explanation through her own interactions within the social world. It is argued by Bozhovich (2004) that "when a person operates with real world objects that were created by human culture throughout history, he [sic] assimilates objectified psychological reality" and this reality as represented through the social and material artifacts and interactions "provides the context for an individual's assimilation of the cultural attainments of past generations" (p. 25). This perspective foregrounds scientific knowledge as not just a cultural construction by society (and as argued in Chap. 1), but also as historically evolving, where this history of knowledge is located in the present moment. This means that for children to access this knowledge system, they need to be oriented to scientific knowledge as an explanatory system for what they experience in everyday life – such as when they see a rainbow in the sky. In this reading of science, scientific thinking is about experiencing their world differently. That is, the children's environment does not change, but their relationship to it does as a result of science teaching. The child who learns about refraction will think very differently about what s/he sees in her or his environment. A rainbow will no longer represent an intangible image that somehow affords looking for the 'end' to yield a pot of gold (as noted in some children's books). In this example, the rainbow is still the same, but the child's relationship to the rainbow has changed, as s/he will think and act differently in relation to the rainbows observed.

Vygotsky (1994) argued that "one should always approach environment from the point of view of the relationship which exists between the child and its environment at a given stage of his [sic] development" (p. 338). What is central here is determining the relation between the child and the social and material environment. This relationship, when expressed from a cultural-historical perspective, takes into account what the child brings to the interaction, and what the activity setting affords for the child. This dialectical view of experiencing the environment means that we can both examine the child's affective attitude as refracted through their previous experience (discussed further below and in Chap. 3), whilst at the same time noting the child's cognitive engagement or orientation to the environment as a source of science learning. Vygotsky referred to this as the interaction between the 'ideal' and 'rudimentary' form. He uses the example of speech to illustrate this concept. For instance, for children to learn to speak a language, regardless of which country they live in, they must be in an environment where people are talking – this representing the ideal form of speech. The child who has rudimentary language, such as an infant, needs to experience the 'ideal form' of language if s/he is to learn to speak. The same argument can be applied to science learning in early childhood centres. If children are to learn to think and act scientifically, they too need to experience a scientific environment – however that is constituted. They will bring, like the language learner, their rudimentary knowledge of how the world works, and through their interactions of the ideal form as an interaction of scientific activity, observation, and explanation with others, will develop higher forms of scientific thinking and acting. Social mediation is central. Vygotsky (1994) stated that "*without social interaction he [sic] can never develop in himself any of the attributes and characteristics which have developed as a result of the historical evolution of all humankind*" (p. 348; original emphasis) including scientific explanations of the world.

A fundamental principle within all of Vygotsky's writings is the view that "*the child's higher psychological functions, his [sic] higher attributes which are specific to humans, originally manifest themselves as forms of the child's collective behaviour, as a form of co-operation with other people, and it is only afterwards that they become the internal individual functions of the child himself* (Vygotsky, 1994, p. 349; original emphasis). This suggests that in science learning, science activity in early childhood centres should be represented in their ideal form, in complete rich and meaningful situations, where children collectively engage in scientific interactions, not as sites for recitation and delivering facts, as is often presented in 'science lessons', but as authentic encounters in the everyday world needing scientific explanation. Here experiencing the preschool environment becomes a scientific orientation, encounter and explanation co-constructed between children and early childhood teachers. Explanation here does not mean 'explaining' but rather is symbolic of an explanatory system for making meaning, and in this particular case, as the cultural knowledge system of science explaining the environment. However, this does not mean that all children will experience the same environment in exactly the same way. Vygotsky's concept of the *social situation of development* is useful here for better understanding why children experience *the same scientific environment differently*.

Vygotsky (1998) introduced the concept of the *social situation of development* through a clinical example from his original research where he discussed how three children from the same family where substance abuse was taking place, experienced their same dysfunctional family differently. What Vygotsky's (1994) research reveals is that the youngest child develops neurotic symptoms, and is simply overwhelmed by the particular environment in which he finds himself. The second youngest child develops an ambivalent attitude to his mother. To Vygotsky's surprise the eldest child (aged 10) who understood that his mother was ill had taken on a special kind of role, of taking on the caregiving for his younger siblings, demonstrating great maturity, seriousness and solicitude. Vygotsky asked "How can one explain why exactly the same environmental conditions exert three different types of influence on these three different children?" (p. 339).

In Vygotsky's case study he argued that the youngest child could not understand what was going on, and therefore felt powerless to affect change, resulting in neurotic symptoms, whilst the eldest child had understanding of the situation, and was therefore able to relate to the same situation that all three children were experiencing in quite a different way. That is, each child brought to their experience of the same environment a different level of psychological development, with the eldest child developing skills way beyond what one might expect of a 10-year-old child. The main point is that each child has their *own special relationship with the environment*, experiencing the same environment differently based on what they bring. We suggest this is the same when a child experiences their scientific environment, each child will already have developed everyday conceptions of their experiences (sometimes referred to as alternative views, see Chap. 1) that they will use to interpret the environment. Different everyday conceptions will yield a very different experience of the same scientific activity and interaction for children.

Bozhovich (2009) in elaborating the social situation of development further, provides an interesting explanation that is worthy of consideration for science education. She argues that "understanding depends (like all other mental processes) on children's affective attitude toward the circumstance affecting them", born out in everyday "observation and analysis of countless pedagogical phenomena" and these observations "attest to the fact that given the same understanding, children often have different attitudes toward one and the same reality, experience it differently, and react to it differently" (p. 68). She goes on to argue that "experiences are products of the reflection of our relationship with surrounding reality" (p. 74). That is, "reflections impels people to act in such a way so as to regulate their interrelationships" and "experiences, once they have taken place and formed a complex system of feelings, affects, and moods, begin to take on significance for people in and of themselves" (p. 74). An example of a child's reflection on their environment and affective attitude in science in early childhood is 'wonder'.

Haddzigeorgiou (2001) puts forward the view that 'wonder' as an emotional quality captures an important relationship between the child and their environment and that this can be pedagogically supported in preschools by teachers. Haddzigeorgiou argues that in building a strong conceptual base through science learning "cannot take place without the establishment of a long-term relationship between the world of science and the child. This relationship can be established only if children are helped to develop certain attitudes towards science"" (p. 64).

A cultural-historical reading of wonder can be conceptualized as an emotional and relational quality that acts as a prism through which the world is experienced by the child. This view of wonder is supportive of Haddzigeorgiou's (2001) comment that "Wonder, in fact, gives things their meaning and reveals their significance" (p. 65). But here, we invest a more dialectical reading by stating that wonder is *not* something that is naturally *within* the child as a scientific way of interacting with the environment, but rather *wonder is socially produced in collective communities*, such as preschool settings, where the ideal form must already be in existence. As with language development occurring as a result of a child being in a language environment, albeit above what the child may fully understand, wonder must also be present within the child's environment as a culturally constructed phenomenon.

In undertaking a cultural-historical reading of the concept of environment from the perspective of early childhood science education, we have noted that the child's experience of the environment demands that an affective relationship not only exist, but as in the case of Haddzigeorgiou (2001) concept of 'wonder' (as a scientific attitude or relationship to the environment), must also be actively developed. We notice this affective relationship of wonder in a study by Siry and Kremer (2011) where Isabella (the teacher) supports two kindergarten children's sense of wonder by actively engaging in their ideas:

Isabelle: *If you want to touch a rainbow, how does I feel?*
Leyla: *If [the rainbow] quickly disappears. And when a child wants to touch it, it quickly disappears so no child can catch it.*
Julia: *I know what Leyla wants to say, when you touch it then you feel nothing at all because then the hand is through it. Because the rainbow I out of nothing.*
Leyla: *So, invisible, right?*
Julia: *No, how could we see the rainbow then?* (p. 648; children are 5 and 6 years old)

An affective relationship between the children and their environment is being built here as the teacher and the children explore rainbows, something that is not only visually appealing, but also intriguing to them. Wonder is being privileged by the teacher as a form of scientific engagement with their environment, as the children explore the different attributes of rainbows through their own physical and imagined interface with the rainbows. Science as a cultural knowledge system is being privileged by the teacher in her encouragement of collective wondering. Here an emotional quality to the children's interactions with their environment is being established by the teacher. *What we see is that the environment is refracted through the lens of scientific wondering.* In Siry and Kremer's (2011) study, wonder was being collectively constructed through particular dialogue, with the following questions asked by the teacher throughout the children's exploration of rainbows:

> What do you see on the picture? ... Have you seen a rainbow before? When and where? ... How does a rainbow arise? ... What does a rainbow feel like? ... Can you stand on a rainbow or use it as a slide? ... What happened when the rainbow isn't there anymore (Siry & Kremer, 2011, p. 654).

Vygotsky (1994) argued that "*Something that is supposed to take shape at the very end of development, somehow influences the very first steps in this development*" (p. 346; original emphasis). That is:

> The greatest characteristic feature of child development is that this development is achieved under particular conditions of interaction with the environment, where this ideal and final form (that form which is going to appear only at the end of the process of development) is

> not only already there in the environment and from the very start in contact with the child, but actually interacts and exerts a real influence on the primary form, on the first steps of the child's development (p. 346).

In early childhood science, a collective sense of wondering represents the ideal form, which children can take up later at a more personal level when experiencing their environment, and through this 'wondering relationship with their environment', a more scientific approach to thinking and learning can be achieved. We see this in the example of Isabella (the teacher) framing the collectively wondering of the children, but also giving conceptual direction to the children's wondering so it leads to a more scientific explanation of rainbows:

Leyla:	Rain and sun.
Julia:	Then it is mixed.
Leyla:	Yes then it is mixed together an there the rainbow comes.
Isabella:	How do you think, the colours arise?
Julia:	The sun has lots of colours in it and then that gets mixed with the rain and then that becomes colours.
Leyla:	Sun and rain. (Siry & Kremer, 2011, p. 653)

Wondering is the affective scientific lens that children use in experiencing their environment. However, only ever wondering without moving conceptually forward means many missed opportunities for science learning feature. We now turn to the work of Tu (2006), who has examined early childhood settings to determine how preschool environments afford science learning for children.

2.3 Preschool Science Environments

Tu (2006) has argued that "as soon as children realize that they can discover things for themselves, their first encounter with science has occurred" (p. 245). In drawing upon Chalufour and Worth (2003, p. 4), Tu states that "wondering, questioning, and formulating ideas and theories" (Tu, p. 245) are part of scientific enquiry into the world surrounding children, and this is a form of 'sciencing'. In a study which sought to examine the opportunities for sciencing in 20 preschool settings in the US, Tu video recorded two consecutive days of morning free play time and analysed both the environment and the activities against two checklists and a coding form. Tu was particularly interested in how preschool settings naturally afford science learning for children. Tu used Neuman's (1972) categories of formal sciencing, informal sciencing and incidental sciencing to examine the environment of the preschool settings.

Here *formal sciencing* refers to specifically planned science activities that are deliberately organized by the teacher, such as providing a cooking activity or introducing a pet into the centre. *Informal sciencing* captures the way in which a teacher might organize a space within the centre for promoting scientific interactions and explorations, such as a science table, or science corner. *Incidental sciencing* refers to interactions that occur between children and the teacher as a result of an occurrence in the centre, such as the weather suddenly changing or a child bringing into the centre a dried seahorse they have found on the weekend, and the teacher in drawing upon scientific concepts elaborates on the child's comments.

In using Neuman's (1972) categories of formal sciencing, informal sciencing and incidental sciencing to analyse the 20 centres, Tu (2006) found that the "activities that the preschool teachers engaged were mostly unrelated to science activities (86.8 %), 4.5 % of the activities were related to formal sciencing, and 8.8 % of the activities were related to informal sciencing" (p. 245). The results show that although half of the preschools had a science area, the teachers mostly spent their time in the art area. Of particular interest is the analysis made by Tu of the materials and equipment for science within the preschool centres. Tu noted that the most common natural materials available to children were plants, seashells, fossils, and pinecones. In addition, vinegar, baking soda, sensory bottles, toad tank, fish tank and tornado bottles were also commonly found in the preschools studied. Tu found that none of these materials were used by the teacher or the children. Interestingly the preschools also had available for children prisms, timers, flower pots, and binoculars, affording a great many possibilities for scientific wondering. None of which were utilized during the data gathering period.

Other opportunities for informal sciencing were reported by Tu (2006) including the provision of a sensory table by 65 % of the centres and a sand or water area in 55 % of centres.

These results would tend to suggest that while there were many opportunities for science learning and a collective sense of wondering about the everyday environment to be created by the preschool teachers, this did not happen. Tu (2006) suggests that "teachers can model with their children a passion for discovery that is common in the world of science. It is acceptable for educators to say "I don't know, why don't we find out together "(p. 251). Tu also suggests that teachers need to exploit the existing science opportunities already available in the centre environments, and argues that if we are "to improve science teaching in the preschool classrooms, teachers need to reflect more on their own practices and utilize the science materials that are available in their environment" (p. 251).

In a study designed specifically for teachers to reflect upon the science opportunities afforded by the preschool environment, Fleer, Gomes, and March (2012) invited teachers to walk with the researchers as they filmed the preschool environment, discussing how children were experiencing science. In using the categories of formal sciencing, informal sciencing and incidental sciencing, and everyday and scientific concept formation (see Chap. 1) to examine the data, they noted:

- As with Tu's findings, science opportunities existed within the constant traditional areas within the preschool (e.g., blocks, sand, water)
- Teachers build science infrastructure into the centre (e.g., light area)
- Teachers and children collectively used science in the everyday life of the centre

A summary of their findings is shown in Table 2.1 and discussed further below.

Table 2.1 Teacher reflections of a scientific experiencing of the preschool environment

Type of science related activity	Sciencing found	Everyday and scientific concept formation found
Formal sciencing	Cooking (Heating, chemical change, change of state of matter)	Composting (decomposition)
Informal sciencing	Overhead projector and coloured blocks (light)	Light area (blocking light, light reflecting and refracting)
		Prism on window sill (refracting light)
		Coloured containers, rainbow stained glass (colour absorption)
		Windmill with coloured blades (white light and spectrum)
		Colour mixing at painting easel (colour absorption)
Science within the constant traditional areas within the preschool	Supporting block building, making concepts explicit for successful building (force)	Water trolley (water wheel – force)
		Sandpit (sand adhering together when wet – force)
		See-saw (force)
Building science infrastructure into the centre		Sensory garden (herbs – use, growth and care)
		Vegetable garden (plant growth and care)
		Flower garden (bulb growth)
Incidental sciencing	Possums in the centre grounds	
	Textured path and chalk (force)	
	Weeding (plant classification in everyday life)	
	Observing birds in the trees (eco-system in centre)	
	Observing flowering of the gum trees in centre (study of plants)	
Using science in everyday life in the centre		Weather watch (Range of concepts)
		Bureau of Meteorology (BOM)
		Rain gauge
		Windmill
		Observing the moon (Earth and beyond)

Adapted from Fleer et al. (2012)

2.3.1 *Science as Part of the Traditional Areas Within the Preschool*

Tu's (2006) analysis of preschools provided evidence of the pervasiveness of opportunities for science learning in early childhood centres, even though teacher-child scientific interactions rarely featured. In the study by Fleer et al. (2012) when teachers were interviewed about the possibilities for science learning within the traditional areas within a preschool, they found not only could the teachers instantly give examples of scientific interactions, but indicated they actively supported a scientific dialogue. An example follows where the teacher stops near the water trolley that is in the outdoor area of the preschool and explains what she commonly observes:

> `They will be pouring` (shows with hands what the children will be doing in the water trolley), `and they will watch the wheels go, so there is a conversation about how the water is able to push the wheel and turn the wheel, and we have a lot of chats, we had a couple of children here yesterday afternoon, and we were having a long chat with, about that` (Fleer et al., 2012, p. 13).

Siry and Kremer (2011) suggest that science opportunities tend to present themselves in relation to what is of interest to children, and that these interests become the resource for supporting the teaching of science in a more informal way. A cultural-historical reading of this would focus on the motive being created, and how motives do not come from within the child, but are developed as a result of children's collective participation in activity settings. In these contexts, not only are children demonstrating a motive for learning, but they are actively encouraged to learn science through teacher-child interactions. Unfortunately, Hedges and Cullen (2005, 2011) have found that in most play-based programs that teachers organise experiences for children as open-ended activities, where the acquisition of content knowledge occurs through osmosis rather than through teaching. Actively focusing on science interactions is generally limited due to teachers being more oriented to other areas of development, than science.

Having a science attitude as part of a teacher's way of interacting with children in the centre means that it is more likely that a motive for science can develop, rather than being observed as a process of osmosis, because as was noted in the study by Fleer et al.'s (2012) the teachers continually and collectively created a sense of scientific wonder and conceptual engagement within the centre. We see this also in other early childhood learning contexts, such as that of Howitt, Upson, and Lewis (2011) who implemented and evaluated forensic science in preschool as scientific inquiry.

2.3.2 *Creating a Science Area – Building Science Infrastructure*

Despite the fact that science areas are common in preschools, the content of these areas tends to focus on the natural environment, and are used mostly to provide interesting objects to explore, but as found by Tu (2006) teachers did not spend time in the area supporting children's wonder, curiosity or conceptual scientific development. However, in the study reported by Fleer et al. (2012) they found that their focus teacher had deliberately set up a physics area within preschool environment (light area) and ensured that it was available all year round for the children. In their study, high levels of teacher-child scientific engagement were noted as occurring from time to time over the 8 week period documented. In the example that follows, the teacher and the 3-year-old child (Henry) are using cellophane blocks which have a wooden frame on the overhead projector (as shown in Fig. 2.1) and are exploring light:

Teacher: Remember you need to lay it flat (pointing to the coloured block) so that that colour (child lays the block flat)... That's it. What colour are you getting now?

Henry: What?

Henry looks to the blocks and then to the wall where the coloured blocks are projecting. He then turns back to the teacher and smiles saying:

Henry: Purple (continuing to smile broadly).

Teacher: It is a purple (nodding at Henry). What about if you try putting one of them on the yellow in the middle? What colour could you put on the yellow one in the middle.

Fig. 2.1 Exploring light

Henry observes the teachers pointed finger, and then takes the block that is in his hand and places it over the yellow block. He then leans over the projector to look closely at the two coloured blocks that are stacked on top of each other.

Teacher: OK. Did you put blue on it or green?

Henry looks to the blocks and also the wall where the colours are being projected. He looks back and forth. Eventually the teacher points to the blocks and says:

Teacher: It is this one, in the middle (tapping with her finger; as Henry looks to her finger and to the wall). What's it done to the colour on the wall?
Teacher: Made it green. It has too. So yellow and blue make green don't they?

Henry smiles and then places two more blocks on top of each other and looks to the what the teacher is doing

Teacher: So what have you put on it?
Henry: Green and red.
Teacher: What colour does that make in the middle?
Henry: Orange.
Teacher: It is a funny kind of green colour on the wall. But it does look orange there (pointing to blocks stacked on the projector) though. So when it's reflected the colour is different.

The teacher then turns to the researcher and says:

The other point about this, is that they are learning that you can't put them up like that (shows block on wooden edge and not flat), that they have to lay them flat. We have had whole conversations about how there is, mirrors and reflections, and the light casting shadows, so a whole lot of learning about light involved in having these (projector and coloured blocks). There is always in this space (pointing to the area) some type of light box, overhead projector, something to do with light and reflection (Fleer et al., 2012, pp. 11–12).

What is special about this example is that the teacher quite deliberately set up a light area as a constant part of the centre. The organization of a specific science focused area to promote high level adult-child dialogue in relation to concepts is rarely featured (see Hedges & Cullen, 2011). The approach adopted, although atypical, provides evidence of explicitly examining scientific concepts in meaningful and iterative ways.

The study by Fleer et al. (2012) found that a *sciencing attitude* was demonstrated through the teacher creating new science infrastructure in the centre along with the traditional areas within the preschool (e.g., block corner) and through making science visible to the children through using it purposefully and in the everyday life of the centre. Their study has shown that a sciencing attitude is something that is important for maximising the science opportunities available to children within early childhood centres.

2.3.3 Using Science in Everyday Life in the Centre

In line with the content of Chap. 1, the study by Fleer et al. (2012) showed that the teacher used science in the everyday life of the centre. We see this took place in many different ways. Two examples are featured, and these examples are drawn from the broader data set (Fleer, 2011–2013):

1. Weather watching
2. Compositing and growing vegetables

2.3.4 Example 1: Weather Watching

The teacher is outside with the researchers and she discusses the extensive weather watching that they do together in the centre, by gesturing to both the natural environment (e.g., clouds), but also to the tools she has in the outdoor area available to the children (e.g., windmill, rain gauge):

> We do a lot of weather watching. We look at the sky. We talk about the sign shining, the wind blowing, makes the 'thingos' go around' (points to the windmill blades), we look at the clouds, so if they are rain clouds coming over, we will talk about the different colours of the different clouds, and what that means, and then we will go in and look at BOM (Bureau of Meteorology). So the children are really familiar with going and looking at BOM, the Bureau of Meteorology web site, at the radar picture, and this is where we live on the map (draws with finger in the air a map symbol, then points), these white and blue spots are the clouds (makes wave movement with arm indicating image on radar), and rain is coming across (motions with hand), they will be over us soon, so let's finish playing outside, and we need to pack up before the rain, then when the rain comes "See, the computer told us the rain was coming, now here it is". So a lot of that sort of thing happens.
>
> BOM helps us plan, what we are going to do, when to get things in, so they don't get wet, but, also the children love it. Certain ones. Not all of them. The ones that ask, we come and sit and we look and talk about it. They just have concepts of computers so well.
>
> We have got the rain gauge (pointing to the gauge). We talk about that occasionally. And we have got the rainbow wind chime (windmill). So there is LOTS of conversation about those, and how the winds pushing it to go round (Fleer, 2011–2013).

In returning to the theoretical arguments put forward at the beginning of this chapter, we see that this teacher had the 'ideal form' of science in the centre. Not only were the artefacts or objects of science available, but the teacher used these tools and the associated scientific concepts for the smooth and effective running of the program in the centre. The children and the teacher collectively studied the clouds, with the purpose of making judgements about if they were rain clouds, and then used the Bureau of Meteorology website to examine the radar in order to determine when it might rain, and what that might mean for playing outside. The teacher is actively orienting the children to their natural environment in a scientific way with a real purpose that is relevant to both the teacher and the children.

The children's experiences of their environment changes as a result of what the teacher points out, discusses, and comments on. Weather watching clearly changes the children's relationship to their environment. The children's experiences of their environment are filtered through a scientific lens. The tools available to the children and the teacher are used to read their environment in scientific ways by measuring rainfall, noticing wind direction, analysing cloud formation, and determining on the radar when it will rain. These events and scientific activities are experienced collectively, with the purpose of deciding if, and when, they can continue to play outside.

2.3.5 *Example 2: Composting and Growing Vegetables*

The teacher is outside and is in close proximity to the vegetable garden. She shows the researchers the garden that is looking quite spent, and discusses the energy cycle from eating fruit, to composting, fertilizing and harvesting vegetables. Although she is interrupted many times during the interview, the intent of her explanation is still evident:

> The vegy [vegetable] garden. You have seen we have compost. They have to divide their food up between compostable and citrus, and rubbish, so what goes into the compost, what goes into the rubbish bin, what we feed the birds, what we feed the possum [to stop him eating the vegetables from the vegetable garden], and we put food down for the ravens, and the cheeky birds [introduced species to Australia], so there is all the composting and then using it.
>
> On Monday we will be digging up the vegy patch, pulling out the things that have had it, digging it over, weeding it, planting some new vegetables. . . [interruption] dig it over, weed it, we talk about what are weeds, and what are plants we want to keep. . . [interruption] so we will plant them, we will water them, I will get another bale of straw, and fertilizer [in addition to using the compost], and then it is all about growing them, I am desperate to get something harvested... (Fleer, 2011–2013).

In having the 'ideal' or authentic form of scientific activity in the child's environment, we see that Bozhovich's (2004) claim that when a child operates with concrete objects that have been created by human culture throughout history, that the child can assimilate not just the scientific concept, but understands how scientific concepts are used to inform actions in everyday life, such as energy transfer as a result of composting all the fruit scraps generated through 'fruit time' in the centre. It is through the teacher-child interactions in collecting the fruit scraps, in putting these into the compost bin, in observing the compositing process, and in using the organic matter for the vegetable garden, that this social and material artifacts and interactions related to energy transfer "provides the context for an individual's assimilation of the cultural attainments of past generations" (Bozhovich, p. 25).

In European heritage communities, most early childhood centres will look similar, and essentially follow the original Froebelian Kindergarten design of 100 plus years ago. Most preschools will be equipped with the traditional child sized furniture and equipment, and be organsised into areas, such as the block corner, the home corner, the puzzle area, the sand pit, painting area, book area, collage area, box construction area, and will have outdoor equipment, such as trestles and balancing beams and a water trolley. This is essentially an imaginary world that really does not represent the child's home or community. These specialized spaces for children's play and learning have remained essentially unchanged. Consciously bringing science into these contexts, either through adding to the traditional areas, such as a physics area, or by using science to run the program, such as using Bureau of Meteorology, afford a very new way of working for early childhood teachers and children. A sciencing attitude affords not just a new way of experiencing the environment for early childhood children, but it gives the possibility for a new way of understanding the environment, as children and teachers collectively draw upon scientific explanations to understand their world.

Imagination in science is clearly an important attribute in learning scientific concepts. In the next chapter we explicitly examine this important area.

References

Bozhovich, L. I. (2004). L. S. Vygotsky's historical and cultural theory and its significance of contemporary studies of the psychology of personality. *Journal of Russian and East European Psychology, 42*(4), 20–34.

Bozhovich, L. I. (2009). The social situation of child development. *Journal of Russian and East European Psychology, 47*(4), 59–86.

Chalufour, I., & Worth, K. (2003). *Discovering nature with young children*. St. Paul, MN: Redleaf Press.

Fleer, M. (2011–2013). *Australian Research Council Discovery Grant*, Unpublished Data. Monash University.

Fleer, M., Gomes, J., & March, S. (2012). A cultural-historical reading of scientific concept formation: Affordances for science learning in preschools. *International Journal of Science Education* (under review).

Haddzigeorgiou, Y. (2001). The role of wonder and 'romance' in early childhood education. *International Journal of Early Years Education, 9*(1), 63–69.

Hedges, H., & Cullen, J. (2005). Subject knowledge in early childhood curriculum and pedagogy: Beliefs and practices. *Contemporary Issues in Early Childhood, 6*(1), 66–79.

Hedges, H., & Cullen, J. (2011). Participatory learning theories: A framework for early childhood pedagogy. *Early Child Development and Care*. doi:10.1080/03004430.2011.597504.

Howitt, C., Upson, E., & Lewis, S. (2011). 'It's a mystery!' A case study of implementing forensic science in preschool as scientific inquiry. *Australasian Journal of Early Childhood, 36*(3), 45–55.

Neuman, D. (1972). Sciencing for young children. In K. R. Baker (Ed.), *Ideas that work with young children*. Washington, DC: National Association for the Education of Young Children.

Siry, C., & Kremer, I. (2011). Children explain the rainbow: Using young children's ideas to guide science curricula. *Journal of Science Education and Technology, 20*, 643–655.

Tu, T. (2006). Preschool science environment: What is available in a preschool classroom? *Early Childhood Education Journal, 33*(4), 245–251. doi:10.1007/s10643-005-0049-8.

Vygotsky, L. S. (1994). The problem of the environment. In R. van der Veer & J. Valsiner (Eds.), *The Vygotsky reader* (pp. 338–354). Oxford, UK: Blackwell.

Vygotsky, L. S. (1998). *The collected works of L. S. Vygotsky, Volume 5: Child psychology* (R. W. Rieber, Ed., & M. J. Hall, Trans.). New York: Plenum.

Chapter 3
Imagination and Its Contributions to Learning in Science

Marilyn Fleer

Abstract This chapter examined the young learner's imaginary play world and explored how this lays an important foundation for scientific thinking. Vygotsky (J Rus East Eur Psychol 42(1):7–97, 2004) argued that 'imagination is not just an idle mental amusement, not merely an activity without consequences in reality, but rather a function essential to life' (p. 13). Imagination becomes the means for broadening a person's experience. Vygotsky (J Rus East Eur Psychol 42(1):7–97, 2004) suggests that humans imagine what they cannot see, conceptualise what they hear from others, and think about what they have not yet experienced. That is, a person 'is not limited to the narrow circle and narrow boundaries of his [sic] own experience but can venture far beyond these boundaries, assimilating, with the help of his imagination someone else's historical or social experience' (Vygotsky, J Rus East Eur Psychol 42(1):7–97, 2004: 17). In this chapter we examined the young child's learning in science through an examination of imagination and creativity in science. Because young learners continually move between reality and imaginary situations in play, it was shown in this chapter that this builds the foundations for thinking with concepts in science. We show through empirical research of science with fairytales how the young learner explores science concepts through their play. The concepts of collective investigations, emotional filtering, duality of emotions and thinking, flickering, and affective imagination are discussed. These are brought together under the concept of perezhevanie.

Keywords Imagination • Creativity • Emotions • Fairytales • Affective imagination • Collective investigations • Emotional filtering • Duality of emotions and thinking • Flickering • Perezhevanie

3.1 Introduction

Matthew (4 years) and his teacher have just been observing a rainbow that had formed on the wall of their preschool. Matthew looks intently at his teacher and says:

```
Matthew:  I saw a rainbow.
Teacher:  When did you see a rainbow Matthew?
Matthew:  In a dream.
Teacher:  In a dream, what a lovely thing.
```


M. Fleer, N. Pramling, *A Cultural-Historical Study of Children Learning Science*, Cultural Studies of Science Education 11, DOI 10.1007/978-94-017-9370-4_3

Matthew:	And I, and it hasn't got any bad things, bad dreams. You were in there, and kids.
Teacher:	So I was there, kids and a rainbow in your dream.
Matthew:	And I had another dream about coming here. Last time it was. But it wasn't that one (pointing to the rainbow on the ceiling).
Teacher:	Which one was it? Is it the one that shines through our prism in the afternoon.
Matthew:	Yeah.
Teacher:	You know the one that shines on the floor and the wall?
Matthew:	No, outside (points out of the window).
Teacher:	Outside. An outside rainbow.
Matthew:	It was it.
Teacher:	Was it in the sky or our wind (moves fingers gesturing windmill action)?
Matthew:	The wind was blowing the rainbow away (gesturing with hands).
Teacher:	Was it?
Matthew:	And I had to go on it (gesturing with hands)
Teacher:	What a beautiful dream (Fleer, 2013).

So how is it that rainbows featured so strongly in Matthew's dreams? Dreaming about rainbows is representative of Matthew's imagination. Mentioning that his dream "hasn't got any bad things" suggests an emotional tone to not just Matthew's imagination, but also his learning in science in this particular preschool.

An emotional tone is also evidence in science generally within our community. Despite the myth of the scientific method, we find prominent scientists use intuition, imagination and emotions in their work. Fox Keller (1983) stated that "Good science cannot proceed without a deep emotional investment…" (p. 197). In her review of the work of Nobel-Laureate Barbara McClintock, she states that McClintock had an ""exceedingly strong feeling" for the oneness of things" (p. 201), where she projected herself inside the microscope joining the chromosomes. She notes that "If you want to really understand about a tumor, you've got to *be* a tumor" (p. 202). McClintock approach was not to position herself outside looking in, but rather she was on the inside being a part of the structures she was seeking to better understand. Through this technique she gained a "feeling for the organism" (p. 201). This approach "both promotes and is promoted by her access to the profound connectivity of all biological forms – of the cell, of the organism, of the ecosystem." (p. 201):

> Her answer is simple. Over and over again, she tell us one must have the time to look, the patience to "hear what the material has to say to you," the openness to "Let it come to you." Above all, one must have "a feeling for the organism." (p. 197).

In learning about how genes can be transposed, where the DNA is influenced by the outside conditions, Barbara McClintock put forward evidence that was in contradiction to the dominant view at the time about genetics. This was gained through a radical way of viewing herself and the material she studied, where emotional connectivity and imagination were clearly evident as a "feeling for the organism"

(Fox Keller, p. 199). This approach has also been named as creativity (see Connery, John-Steniner, & Marjanovic-Shane, 2010).

In this chapter we build upon the previous chapter where we introduced the idea of *wonder as an emotional quality through which children build a particular scientific relationship to their environment*. In that chapter we examine the concepts of *emotional experience* and *environment* for affording science learning in preschools. We draw upon Vygotsky's (2004) dialectical concept of *imagination and creativity* to argue the case that science is a highly imaginative and emotional act, which young children learn within families, communities and preschools. We begin by discussing these terms, followed by examples from an empirical study of children learning science through fairytales (Fleer, 2013). Here we elaborate the concept of *affective imagination for science learning* – where the "feeling for the organism" or the 'feeling for the science concept' is established.

3.2 Imagination and Creativity in Science

How is it that science finds its way into children's imagination? To answer this question, we must first examine what is *imagination and creativity?* Vygotsky (2004) argued that:

> Imagination, as the basis of all creative activity, is an important component of absolutely all aspects of cultural life, enabling artistic, scientific, and technical creation alike. In this sense, absolutely everything around us that was created by the hand of man [sic], the entire world of human culture, as distinct from the world of nature, all of this is the product of human imagination and of creation based on this imagination (pp. 9–10).

In everyday language and conversations, imagination is usually considered as a form of fiction, something that is not real. It is spoken about as though the content were formed solely from within the mind. Creativity also is perceived to have this internal quality, as something that a person invents purely through mental processes, often expressed through the body in some way (e.g., hands, voice, body movement). That is, imagination and creativity are not usually associated with coming from the concrete world, but rather are viewed as fictitious. In using this logic, imagination would appear to be the antithesis of science.

However, Vygotsky (2004) suggests that creative activity is "based on the ability of our brain to combine elements" as an imaginative act (p. 9). It is not always immediately obvious that our cultural world (as distinct from our natural world) has been created through humans as a *result of combining in new ways element taken from our experiences*, or through inheriting the experiences of others. It is argued by Vygotsky that imagination occurs *because* of our experiences. He suggests that the richer our experiences are, the more we have to draw upon, and the more we have at our disposal to combine in new ways. So rather than considering imagination and creativity as something that is unique to the person, as a personal attribute that is genetically transmitted, Vygotsky argued that imagination is acquired through cultural and social interactions within the concrete and social world. This is consistent with contemporary research into creativity, such as that of Ferholt (2010), John-Steiner, Connery, and Marjanovic-Shane (2010) and Lobman (2010) who have also drawn upon Vygotsky's dialectical conception of imagination and creativity.

In line with this theorization, imagination and creativity is deemed to be a product of collectives, rather than individuals, despite the fact that a single person may put forward the new idea, artefact, system, or expression. Patent and copyright laws tend to confirm this individualistic belief (see Wertsch, 1998 for a critique). In science we regularly attribute particular discoveries to individuals (e.g., Boyles Law; Nobel prize), but in the history of science it is clear that individuals stand on the shoulders of past scientists. That is, scientific ideas form because we use past conceptions, drawing from those elements, that we combine in new ways, elements observed through everyday life or through experimentation, or through reading scientific journals or attending conferences and listening to presentations.

Vygotsky (2004) argued that historically, many scientific and technological inventions were formed anonymously and through collective activity over time. For example, "just as electricity is equally present in a storm with deafening thunder and blinding lightening and in the operation of a pocket flashlights, in the same way, creativity is present, in actuality, not only when great historical works are born but also whenever a person imagines, combines, alters, and creates something new, no matter how small a drop in the bucket this new thing appears compared to the works of geniuses" (pp. 10–11). He termed this *collective creativity*.

> When we consider the phenomenon of collective creativity, which combines all these drops of phenomenon of collective creativity that frequently and insignificantly in themselves, we readily understand what an enormous percentage of what has been created by humanity is a product of the anonymous collective creative work of unknown inventors (Vygotsky, 2004, p. 11).

If we apply this logic to young children learning science we can begin to identify imagination and creativity through their play. Vygotsky (2004) argued that "children at play represent examples of the most authentic, truest creativity" (p. 11).

> Everyone knows what an enormous role imitation plays in children's play. A child's play very often is just an echo of what he saw and heard adults do; nevertheless, these elements of this previous experience are never merely reproduced in play in exactly the way they occurred in reality. A child's play is not simply a reproduction of what he [sic] has experienced, but a creative reworking of the impressions he has acquired (p. 11).

In this reading of play and imagination, what the child does in play, is combine prior experiences to create a new concrete situation, one that is focused on the child's own needs and motives.

However, as has been noted by Marjanovic-Shane, Connery, and John-Steiner (2010), traditional thinking and everyday perspectives on concrete situations and fantasy draw a sharp distinction between these terms. For instance, if we consider fairytales, this is an area within early childhood education that is always positioned as pure fantasy. However, Vygotsky (2004) conception of the dialectical relation between fantasy and concrete situation gives a very different reading of fairytales, and this is important for understanding how science can be conceptualised as an imaginative act.

Fig. 3.1 Waiting for hot porridge to cool

In the classic fairytale of Goldilocks and the 3 bears we find bears that live in houses, bears who cook porridge, and bears who sit on chairs and sleep in beds. If we examine this fairytale closely we see that bears, houses, porridge, chairs and beds all exist in reality. However, it is the combination of these things that is unique, imaginary and purely fiction. That is, bears do not live in houses, performing domestic activities, and exhibiting essentially human qualities. Here we see that for young children, imagination builds from concrete situation – from known experiences of living in a house, being part of a family, and waiting for 'hot food to cool' (Fig. 3.1).

Vygotsky (2004) postulated three laws for governing imagination. In the example given, we see that "creative activity of imagination depends directly on the richness and variety of a person's previous experience because this experience provide the material from which the products of fantasy are constructed" (pp. 14–15). That is, a child's experience of waiting for food to cool is a real need, and gives a reason for why Goldilocks could enter the three bears home – no one was there to stop her. Vygotsky suggests that the "richer a person's experience, the richer is the material his [sic] imagination has access to" (p. 15). That is, having experience of eating hot food or cooking gives the experience for children to identify with the bears going on a walk to wait for the porridge to cool, but also the possibility for learning about heating and cooling as scientific concepts. Every imaginative act begins with this accumulation of experience, and in the context of science education, it begins with valuing the experiences and their associated possible alternative everyday conceptions (see Chaps. 1 and 7) as an importance source of scientific concept formation (see Chap. 4). The implication for science education is that children need rich everyday experiences of their world. The richer the experiences, the richer are the possibilities for imaginative and creative thought and action.

The second law put forward by Vygotsky (2004) regarding fantasy and concrete situation centres on how children can appropriate the experiences of others to furnish their imagination. Children do not have to experience in concrete terms a range of different kinds of security systems and locks in order to make sense of how the bears might be able to keep Goldilocks out of their house. Through looking at books or hearing explanations from other children about how they keep their house secured, children can draw upon these vicarious experiences to work imaginatively when 'designing a security system for the three bears'. The linkages between fantasy and the concrete situation are possible through someone else, as a form of social experience. Vygotsky suggested that "In this sense imagination takes on a very important function in human behavior and human development. It becomes the means by which a person's experience is broadened, because he [sic] has to imagine what he has not seen, can conceptualize something from another persons' narration and description of what he himself has never directly experienced" (p. 17). In science, many concepts are not directly observable, and consequently children (and adults) need to imagine these concepts. Children are unlikely to directly see the molecular movement of atoms during the cooling process. Rather children have to imagine the science concepts which help explain how the 3 bears' porridge cools. *Without imagination, thinking with science concepts is difficult.*

The third law for understanding the relations between fantasy and reality put forward by Vygotsky (2004) is emotion. Vygotsky argued that there is a double and mutual dependence between imagination and emotional experience. The *doubleness* can be expressed through the conception that imagination is based on experience *and* experience is based on imagination. The idea is that every experience has an image associated with it; that is, a specific image has a corresponding feeling, an emotional quality. "Emotions thus possess a kind of capacity to select impressions, thoughts, and images that resonate with the mood that possesses us at a particular moment in time" (pp. 17–18). For instance, children who are cooking porridge in anticipation of eating it, waiting for it to cool, will have a different emotional experience to children who do not like porridge, but who nevertheless are expected to eat it. The former creates a positive emotional tone and image for exploring heating and cooing, whilst the latter potentially (if forced to eat the porridge) builds a negative tone and image of the science cooking experience. We see both an external physical expression (disgust at having the eat the porridge; or enthusiastic and joyful anticipation for eating the porridge) and "an internal expression associated with the choice of thoughts, images, and impressions" (p. 18), such as, remembering the eating of porridge at home or imagining the 3 bears walking whilst waiting for their porridge to cool. This duality between the external expression and internal feeling and image bearing state is what is meant by *dual expression of feelings*. Vygotsky suggested that:

> The image of imagination also provides an internal language of our emotion. The emotion selects separate elements from reality and combines them in an association that is determined from within by our mood, and not from without by the logic of the images themselves (p. 18).

In looking at the porridge cooling, the child whose duality of emotion and experience is positive potentially imagines the feeling state of the bears, wanting to cool their porridge, thinking about the cooling process (particularly when supported to consciously consider the concept of cooling by the teacher), and potentially *imagining* how this might occur. The emotional tone for the science cooking experience is positive, the anticipation of eating the porridge is foregrounded, making the cooling process more urgent, and the learning situation has an emotional quality that makes science a positive event. All forms of imagination include an affective tone or quality. This mutuality of emotions and imagination is captured in the concept of *affective imagination* and is centrally important in understanding how very young children experience science.

In returning to the example of Matthew's dream of the rainbow, we also see an emotional quality to his science learning as represented in his dream, as something positively experienced. Vygotsky (2004) in citing Ribot says:

> These types of associations are very often present in dreams or day-dreams, that is, in states of mind in which the imagination has free rein and works at random, any which way (not dated).

It can therefore be argued that the affective imagining of rainbows by Matthew characterizes how both imagination and the concrete situation give meaning to each other in science learning.

In order to fully appreciate the relations between imagination and concrete situation, we must also give thought to how combining new elements in the process of imagining something new, must be substantially new, if imagination is to turn into a concrete situation. It is through the process of realising images and thoughts into concrete situation or constructions, such as occurs in the development of machines, the cycle of imagination becomes complete, as a creative act. We see this when children anticipate the eating of porridge, wishing for the hot porridge to cool quickly, or more concretely in the story of Goldilocks and the 3 bears, discussing how they can help the bears cool down their porridge by inventing a 'cooling down machine'. What is imagined becomes concrete and tangible as a new creation, as a cooling down machine. Here it is possible to see how "the intellectual and the emotional – are equally necessary for an act of creation" (Vygotsky, 2004, p. 21). The invention of the cooling down machine as a concrete creation now has a role in the story and in the play of the children, influencing reality. Similarly, imagining heat transference as a cooling process (perhaps not at the molecular level for young children), also influences reality because children have concrete actions they can now take in everyday life, such as stirring the porridge or putting a metal spoon in the porridge to aid cooling. Actions are changed due to the new meaning given, and here we see scientific imagining being foregrounded.

To illustrate these concepts more concretely, we now turn to a case example of using fairytales for science learning where affective imagination is explicitly featured.

3.3 Case Example: Learning Science Through Goldilocks and the 3 Bears

It is well understood that imagination and creativity are featured in children's play (Connery et al., 2010; Ferholt & Lecusay, 2010; Holzman, 2009; Vygotsky, 2004). How teachers draw upon play to further science learning has not always been well articulated. Rather what dominates the literature is conceptual understandings in science – notably conceptual change. This well established body of literature suggests that young children can learn many scientific concepts at a very early age (Eshach, 2011; Fleer, 2009a, 2009b; Goulart & Roth, 2010), such as, astronomy (Hannust & Kikas, 2007; Robbins, 2003; Sharp, 1995), electricity (Fleer, 1990, 1991; Fleer & Beasley, 1991), food (Cumming, 2003), digestion (Martins Teixeira, 2000), natural science (Keleman, 1999a; Ravanis & Bagakis, 1998; Shepardson, 2002, Venville, 2004), force (Hadzigeorgiou, 2002), matter (Krnel, Watson, & Glazar, 2005), as well as engage in co-constructing science with teachers (Goulart & Roth, 2010), engaging in epistemological reasoning (Pramling & Pramling Sameulsson, 2001; Tytler & Peterson, 2003), and teleological thinking (Keleman, 1999b). As noted in Chap. 1, much of the empirical work has been conceptualized from a constructivist perspective, with exceptions emerging in recent years (such as, Goulart, Pramling, Robbins, Roth) where more of a cultural-historical orientation has framed the research. The case study that follows drew upon a cultural-historical view of learning science.

One of the defining features of preschools is the existence of play-based programs. A play-based program is distinct from how learning is generally organised in both primary and secondary schools. The preschool from which the case study is drawn, is structured so that group learning usually occurs through both play periods and two 30 min sessions of teacher organised group time, where stories, role play, singing games, and the like are featured. Mostly children make choices about what they will do from a range of activities and infrastructure during the free play periods. The group sessions are usually organised by the teacher and all children usually participate in these sessions. The framework for science learning was the fairytale of Goldilocks and the 3 bears. Five dimensions were featured, and they are discussed in turn.

3.3.1 *Collective Investigations and Narratives*

The organizational structure of the preschool featured the telling and re-telling of Goldilocks and the 3 bears, followed by using the available props for role-playing the story. In particular an *Imagination Table* with bowls, bears, beds, etc., for role-playing was set up for the children, where an iPad allowed the children to capture pictures of their play (see Fig. 3.2). Also available were experiences which gave a more scientific reading of what was being introduced to the children in group time – that is, the teacher set up over a period of 8 weeks many opportunities to cook and eat porridge and to design and make a cooling down machine, something that emerged from the children as a way of helping the 3 bears to quickly cool their porridge so that there was no need for the bears to leave their house. See Fig. 3.3.

Central to the collective imaginary situation that emerged was a series of collective investigations. Through the telling of the fairytale, where the children identified with the bears, where they sought to assist the bears with cooling their porridge (see Table 3.1). This was a highly pertinent narrative, because, as mentioned previously, children regularly wait for hot food to cool before they can eat it.

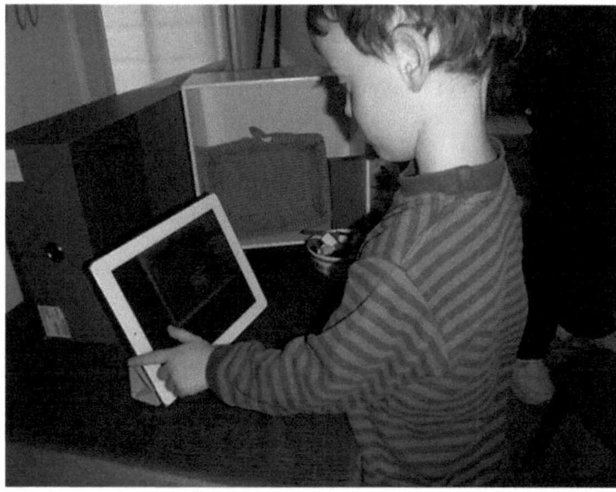

Fig. 3.2 Imagination table with iPad

Fig. 3.3 Cooking porridge – consciously considering heating and cooling

Through cooking porridge with the children, the teacher re-created an everyday situation common in all families, but also specific to what was central to the story of Goldilocks and the 3 bears. The teacher generated a scientific narrative as part of cooking porridge.

Table 3.1 Collective scientific investigations

Concept	Emotionality in fairytales	Emotionality in scientific and technological learning
Collective investigations and narratives	Children *want* to identify with the hero of the story, wishing to assist the hero, and through this, they *together re-enact the ideal moral response to the given situation, along with all of the associated risks, in reaching the final victory.*	*Collective scientific investigations* *Children collectively* develop a consciousness of scientific and technological *concepts and emotionality* by working together with other children to solve the problem.
	Children imagine the feeling state of the fairytale characters, and empathise and want to help the characters to solve the collective problem.	In a *scientific narrative*, children empathise and want to help the characters to solve the collective scientific and technological problem.

See Fleer (2013)

3.3.2 *Affective Imagination*

In the case example, the children not only had experiences of role-playing Goldilocks and the 3 bears with the teacher, but they also actively re-created the story during free play time where they used a scientific narrative, as occurred when the children took props relevant to the fairytale and role-played cooking and cooling porridge.

Jason (3 years) is at the 3 bears table. He has taken to the table a bowl of small cut straws and is pouring these into the 2 equal sized bowls that are at the table. One larger bowl also stands on the table. Jason pours the straw pieces back into the basket, and then turns to the research assistant Shukla and asks:

What can I get for you today? Shukla says she would like something.

Jason: Porridge?
Shukla: Yes. I'd like porridge.
Jason: Porridge.

Jason takes the small basket of sticks, shakes them around as he says:
Jason: But, I'm going to put it into the microwave, because it get's very hot.
Shukla: OK. Is it too hot?
Jason: Yes (shaking the basket of cut drinking straws). When I put it in this bowl (about to pour the cut straws into the bowl). Do you want it in this middle sized bowl or the big one, 'cause we don't do middle sized ones (shaking his head). Do you want a little one (correcting himself) or a big one, 'cause we don't do middle sized ones?

Table 3.2 Affective imagination in science

Concept	Emotionality in fairytales	Emotionality in scientific and technological learning
Affective imagination or emotional imagination	Through the re-enactment of fairytales, children gain a sense of the main character's actions in role-play, whilst clarifying their own feeling state because the story plot is mirrored in the acted out actions of the children.	Through role-play of scientific narratives and learning, the *children collectively* begin to anticipate the results of each others' actions in the play, begin to anticipate their own actions, including image-bearing dramatization, verbal descriptions, prop use and transformation, and importantly, the scientific solutions created through the support of the teacher.
Zaporozhets (2002) shows that through emotional and cognitive participation in fairytales that children reach "the ideal plane of *emotional imagination*" (p. 58).	Children are not *"enacting the story,* but *really living in it"* (El'koninova, 2002, p. 45).	

Through role-play children not only begin to use the language found in the fairytale (e.g., big, little and middle sized bowls of porridge) but also use rudimentary scientific language, such as, the use of 'hot' in an everyday context of pretending to serve porridge. Here children make conscious the concept of 'heating and cooling'. This play represents the beginnings of a scientific narrative (remembering these children are 3–4 years of age) where they are imagining 'hot food' in the same way as they will need to do later to imagine molecular movement in order to understand scientific explanations of heating and cooling (Table 3.2).

3.3.3 Being in and out of Imaginary Situations – Flickering

During play children move in and out of imaginary situations. That is, they are part of the role-playing reliving the story, but when the play does not progress as expected, then children slip out of the imaginary situation and direct the play from outside of the play. For example, we would see this when children are in role acting out being the bear cooking the porridge, but if one of the children deviates from the story line by saying "I can eat up all the porridge because it is just right", then the other children would coach them back in, by saying 'No, you have to say "the porridge is too hot"'. We also find children direct the play from within the imaginary situation, as is evident when a child might say in character and say: "But then Goldilocks comes along and she says 'the porridge is *too hot*'" (in an exaggerated tone). In this example the children are concurrently in the imaginary situation and the real world. Concurrently being in the imaginary situation and the real world helps children interrogate the concepts as they play, allowing for a more conscious response to a play or learning situation. Like with the approach taken by Barbara McClintock in her genetic work, the children are inside of the situation feeling the heating and cooling in their play. This is an important aspect of everyday and scientific concept formation, where a conscious exploration of a concept allows children to build deeper understandings in science. In role-playing heating or cooling, children have to exaggerate or explicitly show the concepts concretely or symbolically in their play, if other children are to understand and engage in the imaginary situation forming. In so doing, children make the concepts conscious, and thereby consciously explore the concept.

Flickering in and out of the imaginary situation supports children to build their imagination, as they actively enter into an imaginary world. Why this is important in science is that many aspects of science are not directly observable by children. Many of the scientific concepts, such as magnetism, gravity, molecular movement, the Earth's rotation around the sun, have to be imagined. Imagining scientific explanations for not directly observable phenomena is an important dimension of learning in science for preschool children. Yet this dimension of science is not always acknowledged. This flickering is represented in Fig. 3.4 (Adapted from Fleer, 2013) and also in Table 3.3.

Fig. 3.4 Flickering between the imaginary world and the concrete world

Table 3.3 Imagining non-observable concepts in science

Concept	Emotionality in fairytales	Emotionality in scientific and technological learning
Flickering Children flicker between the concrete and imaginary worlds.	In fairytales, children begin to separate out the imaginary world from the concrete world, and find themselves in the borderline between these worlds.	It is the border of the imaginary world and the concrete world that creates a dialectical relation and emotional tension that promotes scientific conceptual development, which helps children imagine scientific explanations not easily observable.

3.3.4 Duality of Emotions and Thinking

Fairytales are full of anticipation, where emotional responses usually feature – such as feeling frightened or excited – even thought the outcome of the storyline is well known to children. In identifying with the characters in the fairytale, either in the re-telling or in the role-play, children live through the emotions of the story. Identifying with the character, wishing to help them to solve the problem, are laden with emotions. Vygotsky (1966) argued that in play children can feel two things concurrently. They can feel the joy of playing, while also feeling the emotions of the characters – such as being frightened. This is relevant to science because children can also experience an emotional response to science learning. That is, they may feel happy exploring the science problem while feeling anxious about needing to solve the scientific challenge quickly for the role-play. The duality of emotions has also been noted in McClintock's work, when she discusses her delight for contradiction and surprise during the post-World War II period where the effects of radiation on flies (Drosophila) was being investigated:

> "It turned out that the flies that had been under constant radiation were more vigorous than those that were standard. Well, it was hilarious; it was absolutely against everything that had been thought about earlier. I thought it was terribly funny; I was utterly delighted" (p. 198).

In scientific investigations, children's feeling state becomes connected with the learning as they anticipate *finding a solution*. Through consciously considering feeling states in science, emotions become intellectualized, generalized, and anticipatory, while cognitive processes acquire an affective dimension, performing a special role in meaning discrimination and meaning formation (e.g., gut feeling this is going to work). The duality of external expression, and internal feelings and images, occurs simultaneously. Imagination is based on these dual experiences/images but both become emotionally charged in the process of being experienced, imaged and created (Table 3.4).

Table 3.4 Emotions and scientific thinking

Concept	Emotionality in fairytales	Emotionality in scientific and technological learning
Dual role of emotions in thinking	Children must be inside of the plot living the story, and outside of the plot as a real person. El'koninova (2002) argues that a child must "gropingly look for a "territory" where this is possible" (p. 41).	Children feeling happy enacting or exploring a science narrative with others, but also feeling excited or curious by learning new things and solving scientific and technological problems in order to scientifically help the characters in the narrative.
	Feeling happy in role-play, but also feeling frightened when pretending to be Goldilocks seeing the 3 bears.	

Barbara McClintock has also demonstrated these connections between emotions and thinking, when she became intrigued by the way Tibetan Buddhists could control their body temperature. McClintock's wonder and "feeling for the organism approach" led her to experiment with biofeedback, where "she began to feel a sense of what it took" (p. 200):

> "I was so startled by their method of training and by its results that I figured we were limiting ourselves by using what we call the scientific method"... "We are scientists, and we know nothing basically about controlling our body temperature. [But] the Tibetans learn to live nothing but a tiny cotton jacket. They're out there cold winters and not summers, and they have been through the learning process, they have to take certain tests. One of the tests is to take a wet blanket, put it over them, and dry that blanket in the coldest weather. And they dry it." (p. 200).

3.3.5 *Emotional Filtering*

In the case example of fairytales, the teacher used a lot of emotional filtering. That is, she regularly emotionally charged events and actions, which children responded to positively. For instance, in the example that follows the teacher emotionally charges the concept of cooling by foregrounding the word 'hot':

> Four children are sitting or standing around a table which has a large pile of Lego pieces in the centre. The teacher is seated at the table. She begins a discussion about porridge making so that she can discuss the idea of designing some sort of device for cooling down the porridge:

Teacher: Remember what the 3 bears cooked and ate for breakfast?
Child 1: Porridge.
Teacher: Yum. Do you remember how to make porridge?

```
Child 2:  Yeah (other children nod in agreement).
Teacher:  How did we make it?
Child 1:  With some milk.
Teacher:  Milk. Yes. And?
Child 3:  Then you put it into the microwave.
Teacher:  And what did the microwave do to make porridge?
Child 2:  Warm it up.
Child 3:  Make porridge.
Teacher:  Warmed it up, or cooked it?
Child 2:  Cooked it.
Teacher:  What was it like when it came out of the microwave?
Child 3:  Hot.
Child 2:  Hot.
Teacher:  A little bit hot, or very, very, very, very, very,
          very, very, very, very, very, hot.
Child 3:  Very hot.
Child 2:  Very hot.
Teacher:  It was nearly boiling.
Child 3:  It was.
Teacher:  Why did it nearly have to be boiling?
Child 3:  Because it was in there for a long time.
Teacher:  So we have got boiling hot porridge. So can we eat
          it when it's boiling hot?
Child 3:  No (all children shake their heads). My grandmother
          only eats porridge when its cold.
Teacher:  Does she wait until its got cold before she eats it?
Child 3:  Yeah and puts yoghurt in it.
```

This is common practice in early childhood centres, where teachers regularly highlight something through an enthusiastic response or exaggeration. In this case, the teacher does this specifically in relation to a science concept. This draws the children's attention to the concept, and to the scientific challenge that presents itself. By making concepts conscious to children through emotional filtering, teachers are able to work informally in preschool settings, drawing out science in the everyday life of the program (Chap. 2), as well as create science through events that children find emotionally and intellectually interesting, such as the Goldilocks and the 3 bears. See Table 3.5.

Table 3.5 Emotional filtering in science

Concept	Emotionality in fairytales	Emotionality in scientific and technological learning
Emotional filtering *Emotional filtering* is "where kindergarten teachers attribute emotional significance to events" (Iakovela, 2003, p. 93).	Teachers emotionally charge events, actions and objects which focus the children's attention, thinking and feeling state.	Teachers help children in knowing what is noteworthy to pay attention to in science learning. What should they notice or look for? The gesturing of teachers is usually accompanied by expressive sounds and surprised or interested facial expressions.

3.3.6 Wholeness Approach to Science Learning in Preschools

Through an emotional connectivity, that the teacher foregrounds, children explore the wholeness of science. That is, they do not learn discrete parts of science (i.e., single concepts out of context), but rather are making meaning of science in everyday life, as we saw in chapter 3. However, the teacher's emotional filtering, can also be used effectively in creating imaginary situations, such as occurs in fairytales, where the teacher ensures that the imaginary situation affords scientific investigating.

What is important here is how the teacher emotionally filters science to the children. As Fox Keller (1983) states in her explanation of McClintock's approach to genetics, "without an awareness of oneness of things, science can give us at most only nature-in-pieces; more often it gives us only pieces of nature." (p. 201).

Affective imagination is foregrounded in the case example discussed. But as has been alluded to throughout this case example, Nobel-Laureate scientists too, work in ways that do not follow the mythical scientific method. As McClintock states "So you work with so-called scientific methods to put it into their frame after you know." (p. 200). But to get there, McClintock argues that you have to spend lots of time getting to know what you are seeking to study – in her case plants:

> One must understand "How it grows, understand its parts, understand when something is going wrong with it. [An organism] isn't just a piece of plastic, it's something that is constantly being affected by the environment, constantly showing attributes or disabilities in its grown. You have to be aware of all of that . . ." You need to know those plants well enough so that if anything changes, . . . you [can] look at the plant and right away you know what this damage you see is from – something that scraped across it or something that bit it or something that the wind did. *"You need to have a feeling for every individual plant"* (p. 197; our emphasis).

Fox Keller (1983) states that it is that emotional investment that provides the "motivating force for the endless hours of intense, often grueling, labor" (pp. 197–198). In the case example, the teacher worked with the 3 year old children for 8 weeks exploring heating and cooling. Whilst this was not grueling, it was an investment in time and emotional energy by the children and the teacher, as they role-played the 3 bears, experienced the cooking of porridge, invented a cooling down machine, and created their own slowmation of cooking porridge in the context of the fairy tale of Goldilocks and the 3 bears. As McClintock states "I don't feel I really know the story if I don't watch the plant all the way along. So I know every plant in the field. I know them intimately, and I find it a great pleasure to know them." (p. 197).

A wholeness approach to science teaching in preschools foregrounds, the following characteristics for imagination, creativity and emotions:

1. *Collective investigations and narratives*
2. *Affective imagination*
3. *Being in and out of imaginary situations – flickering*
4. *Duality of emotions and thinking*
5. *Emotional filtering*

3.4 *Perezhevanie* as an Explanation of Quality Early Childhood Science Learning

In concluding this chapter, we theorise further the relations between imagination, concept formation and emotions, by drawing upon Vygotsky's concept of *perezhivanie*. According to Veresov (2012) this word does not translate well into English. We have chosen to use the Russian term in our discussion of emotions and imagination in early childhood science education, because we believe this term captures and helps us to better understand how preschool children experience and learn science as they interact with their social and material world. Central to this concept is the idea that emotions, imagination and concept formation must be conceptualized in unity. That is, they cannot be separated out, as is often the case in early childhood science education where only the learning dimensions are discussed. All children, but especially young children, relate to their social and material world emotionally. Young children are still learning to regulate their emotions, and this means they are not always in a position to consciously think about their feeling state – this is after all something they develop throughout the early childhood period. To conceptualise science education, as an affective and imaginary experience, means that the central concepts discussed in this chapter need to be brought together. Perezhivanie captures this unity. As Vygotsky (1994) states:

> *An emotional experience [perezhivanie] is a unit where, on the one hand, in an indivisible state, the environment is represented, i.e.,* that which is being experienced – an emotional experience *[perezhivanie]* is always related to something which is found outside the person – *and on the other hand what is represented is how I, myself, am experiencing this*, i.e.,, all the personal characteristics and all the environmental characteristics are represented in an emotional experience *[perezhivanie];* (Vygotsky, 1994, p. 341; Original emphasis).

Consequently, we conceptualise science education as an indivisible unity of what the child brings to the activity setting in the preschool, the situational characteristics that are created by the teacher, as well as how these events are emotionally and conceptually experienced by the child. Together these represent the emotional experience or perezhivanie of the child's social situation of development. Veresov (2012) in line with Vygotsky (1994) has argued that perezhivanie is the prism through which both the individual and the socio-cultural environment is experienced, and together they represent the unit of human consciousness as a central force for human development. Consequently, perezhivanie is a cultural form of experiencing the scientific environment, and because play is the leading activity within the preschool period (Vygotsky, 1966), imagination and creativity are featured as part of perezhivanie, and together emotions, cognition and imagination become central dimensions for science teaching and learning.

References

Cummings, J. (2003). Do runner beans really make you run fast? Young children learning about science-related food concepts in informal settings. *Research in Science Education, 33*(4), 483–502.

Connery, M. C., John-Steiner, V. P., & Marjanovic-Shane, A. (Eds.). (2010). *Vygotsky and creativity: A cultural-historical approach to play, meaning making and the arts* (pp. 163–180). New York: Peter Lang.

El'koninova, L. I. (2002, March–April). The object orientation of children's play in the context of understanding imaginary space – Time in play and in stories. *Journal of Russian and East European Psychology, 39*(2), 30–51.

Eschach, H. (2011). Science for young children: A new frontier for science education. *Journal of Science Education and Technology, 20*, 435–443.

Ferholt, B. (2010). A synthetic-analytic method for the study of perezhivanie: Vygotsky's literary analysis applied to playworlds. In M. C. Connery, V. P. John-Steiner, & A. Marjanovic-Shane (Eds.), *Vygotsky and creativity: A cultural-historical approach to play, meaning making and the arts* (pp. 163–180). New York: Peter Lang.

Ferholt, B., & Lecusay, R. (2010). Adult and child development in the Zone of Proximal Development: Socratic dialogue in a playworld. *Mind, Culture, and Activity, 17*, 59–83.

Fleer, M. (1990). Scaffolding conceptual change in early childhood. *Research in Science Education, 20*, 96–103.

Fleer, M. (1991). Socially constructed learning in early childhood science education. *Research in Science Education, 21*, 96–103.

Fleer, M. (2009a). Understanding the dialectical relations between everyday concepts and scientific concepts within play-based programs. *Research in Science Education, 39*, 281–306.

Fleer, M. (2009b). Supporting scientific conceptual consciousness or learning in 'a roundabout way' in play-based contexts. *International Journal of Science Education, 31*(8), 1069–1089.

Fleer, M. (2013). *Affective imagination* in science education: Determining the emotional nature of scientific and technological learning of young children. *Research in Science Education*. doi:10.1007/s11165-012-9344-8. Published on line Jan 2013.

Fleer, M., & Beasley, W. (1991). A study of conceptual development in early childhood. *Research in Science Education, 21*, 114–123.

Fox Keller, E. (1983). *A feeling for the organism: Life and work of Barbara McClintock*. New York: Freeman.

Goulart, M. I. M., & Roth, M. (2010). Engaging young children in collective curriculum design. *Cultural Studies of Science Education, 5*, 533–562.

Hadzigeorgiou, Y. (2002). A study of the development of the concept of mechanical stability in preschool children. *Research in Science Education, 32*, 373–391.

Hannust, T., & Kikas, E. (2007). Children's knowledge of astronomy and its change in the course of learning. *Early Childhood Research Quarterly, 22*, 89–104.

Holzman, L. (2009). *Vygotsky at work and play*. London: Routledge.

Iakovela, E. L. (2003). Emotional mechanisms underlying personal and creative development. *Journal of Russian and East European Psychology, 41*(6), 92–100.

John-Steiner, V., Connery, M. C., & Marjanovic-Shane, A. (2010). Dancing with the muses: An cultural-historical approach to play, meaning making and creativity. In M. C. Connery, V. P. John-Steiner, & A. Marjanovic-Shane (Eds.), *Vygotsky and creativity: A cultural-historical approach to play, meaning making and the arts* (pp. 3–16). New York: Peter Lang.

Keleman, D. (1999a). Why are rocks pointy? Children's preference for teleological explanations of the natural world. *Developmental Psychology, 35*(6), 1440–1452.

Keleman, D. (1999b). The scope of teleological thinking in preschool children. *Cognition, 70*, 241–272.

Krnel, D., Watson, R., & Glazar, S. A. (2005). The development of the concept of 'matter': A cross-age study of how children describe materials. *International Journal of Science Education, 27*(3), 367–383.

Lobman, C. (2010). Creating developmental moments: Teaching and learning as creative activities. In M. C. Connery, V. P. John-Steiner, & A. Marjanovic-Shane (Eds.), *Vygotsky and creativity: A cultural-historical approach to play, meaning making and the arts* (pp. 199–214). New York: Peter Lang.

Marjanovic-Shane, A., Connery, M. C., & John-Steiner, V. (2010). A cultural-historical approach to creative education. In M. C. Connery, V. P. John-Steiner, & A. Marjanovic-Shane (Eds.), *Vygotsky and creativity: A cultural-historical approach to play, meaning making and the arts* (pp. 215–232). New York: Peter Lang.

Martins Teizeira, F. (2000). What happens to the food we eat? Children's concepts of the structure and function of the digestive system. *International Journal of Science Education, 22*(5), 507–520.

Pramling, N., & Sameulsson, P. (2001). "It is floating 'cause there is a hole'": A young child's experience of natural science. *Early Years, 21*(2), 139–149.

Ravanis, K., & Bagakis, G. (1998). Science education in kindergarten: Sociocognitive perspective. *International Journal of Early Years Education, 6*(3), 315–327.

Robbins, J. (2003). The more he looked inside the more Piglet wasn't there: What adopting a sociocultural perspective can help us see. *Australian Journal of Early Childhood, 28*(2), 1–7.

Sharp, J. (1995). Children's astronomy: Implications for curriculum development at Key Stage 1 and the future of infant science in England and Wales. *International Journal of Early Years Education, 3*(3), 17–49.

Shepardson, D. P. (2002). Bugs, butterflies, and spiders: Children's understandings about insects. *International Journal of Science Education, 24*(6), 627–643.

Tytler, R., & Peterson, S. (2003). Tracing young children's scientific reasoning. *Research in Science Education, 33*(4), 433–466.

Venville, G. (2004). Young children learning about living things: A case study of conceptual change from ontological and social perspectives. *Journal of Research in Science Teaching, 41*(5), 449–480.

Veresov, N. (2012). *Perezhivanie and cultural development: A key which opens the door?* Unpublished paper presented at Monash University, Peninsula campus.

Vygotsky, L. S. (1966). Play and its role in the mental development of the child. *Voprosy Psikhologii, 12*(6), 62–76.

Vygotsky, L. S. (1994). The problem of the environment. In J. Valsiner & R. vander Veer (Eds.), *The Vygotsky reader* (pp. 347–348). Oxford, UK: Blackwell.

Vygotsky, L. S. (2004). Imagination and creativity in childhood. *Journal of Russian and East European Psychology, 42*(1), 7–97.

Wertsch, J. V. (1998). *Mind as action*. New York: Oxford University Press.

Zaporozhets, A. V. (2002). Toward the question of the genesis, function, and structure of emotional processes in the child. *Journal of Russian and East European Psychology, 40*(2), 45–66.

Chapter 4
Theoretical and Conceptual Insights – The Young Learner in Science

Marilyn Fleer

Abstract In this chapter a brief discussion of the concepts outlined in the first section of the book are revisited and brought together to give new insights into concept formation for the very young learner: Critiquing the historical legacy of science education research in relation to a set of 'taken for granted' assumptions; foregrounding the everyday concepts that young children hold, rather than being seen as getting in the way of scientific learning; framing science learning as a dialectical relations between scientific concepts and everyday; and conceptualizing science knowledge as dynamic. In drawing together these four principles, this chapter theorises a new set of assumptions for shaping the development of a cultural-historical view of learning in science for young children.

Keywords Historical legacy • Dialectical relations • Science knowledge as dynamic

4.1 Introduction

In this first section of the book we have examined the theoretical foundations for learning science in early childhood. In particular, we looked at the empirical and theoretical literature in order to gain understandings about how others have conceptualized children's learning in science. Because we were interested in a cultural-historical framework for preschool children's conceptual thinking in science, we specifically examined this literature, noting that only a small pool of material was available to us. However, in extending our analysis through drawing upon the collected works of Vygotsky, we were able to give a new kind of reading for science learning.

4.2 Principles for Science Learning

In this section of the book we sought to move away from the traditional empirical studies that have generally argued that everyday concepts get in the way of scientific concepts, and have instead been guided by Vygotsky's (1987) work on the dialectical relations between everyday concepts and scientific concepts. Here we argued

M. Fleer, N. Pramling, *A Cultural-Historical Study of Children Learning Science*, Cultural Studies of Science Education 11, DOI 10.1007/978-94-017-9370-4_4

strongly that children need everyday experiences of science in order to name and work with new meanings of their environment – that is, everyday experiences and concepts do not get in the way, but rather provide the rich tapestry from which scientific understandings can form. We argued that through a consciousness of concepts, on the part of both the teacher and the children, new meanings of the environment could be made. That is, the children's environments do not change, but rather how they think about their environment does as a result of science teaching. From this perspective it can be argued that a new way of thinking about science learning in early childhood education is needed.

We also examined a different reading of preschool science education. We focused on imagination, creativity and emotions in science. When we use cultural-historical theory for framing our understanding of teaching and learning, where the dialectal relations between everyday concepts and scientific concepts are the focus of our attention, and imagination and creativity are considered alongside of how the learner feels, a whole new orientation to early childhood science education emerges. We noted this as being important for both re-theorising our conceptions of early childhood science education, but also we can now see that a new set of principles to guide our day-to-day work in preschool centre is needed.

As such, we present the following principles to inform a new conception of early childhood science education:

Principle 1: *We must recognize and scrutinise the historical legacy of science education research because this frames what we see and think, contributing to a set of 'taken for granted' assumptions that may no longer be helpful*

Principle 2: *A cultural-historical reading of science education means that we see everyday concepts of children as central and not as getting in the way of scientific learning*

Principle 3: *The dialectical relations of learning science means we must develop both scientific concepts and everyday concepts if conceptual development is to occur – this is different to conceptual change*

Principle 4: *Science knowledge is not static, therefore why would we assume one explanation of the material and natural world is the 'right' one.*

These four principles can be drawn together to theorise a new set of assumptions for shaping the development of a cultural-historical view of learning in science for young children (see Chap. 13). An overview of the key assumptions and theoretical drivers for informing a cultural-historical view of early childhood science education are shown in Table 4.1.

In this first section of the book we also studied how imagination, creativity and emotions shaped and was shaped by science education. In particular we put forward a wholeness perspective, where five characteristics featured:

1. *Collective investigations and narratives*
2. *Affective imagination*
3. *Being in and out of imaginary situations – flickering*
4. *Duality of emotions and thinking*
5. *Emotional filtering*

A wholeness approach to science teaching in preschools foregrounds these characteristics, suggesting that the child, the concept and the social and material world cannot be separated from each other. They each give meaning to science teaching, and collectively ensure they make learning meaningful for the young learner. In Table 4.2 we bring together these characteristics and argue that for young children preschool science needs to encompass affective imagination as a central dimension in teaching and learning.

The principles and concepts outlined in Tables 4.1 and 4.2 begin to capture the complexity of what matters in the teaching of science to very young children. In the next section we progress these key assumptions and theoretical drivers further. There we examine the empirical literature on early childhood science learning across a range of countries. We specifically explore these studies from the theoretical framework introduced in this first section. What we seek to do is find out how children learn science in play-based settings across cultures.

Table 4.1 Key assumptions and theoretical drivers for informing a cultural-historical view of early childhood science education

	Everyday concepts	Scientific concepts	The dialectical relations between everyday and scientific concepts
1. *The historical legacy of science education research:* Problematising the dualism between Children's Science and scientific concepts.	Everyday concepts are important in conceptual development, they do not get in the way of children's learning.	Scientific concepts are formed as a result of a child's 'extraordinary efforts' in his/her own thought processes and not through assimilation or memorization.	Conceptual development is immeasurably more complex and positive than the idea of cognitive conflict.
2. *The dialectical relations of learning science:* Everyday concepts can only be understood in relation to scientific concepts. Scientific concepts can only understood in relation to everyday concepts.	Everyday concepts build broadly and intuitively, capturing the dynamic flux, ebb and flow of a child's interactions with their material and social world.	Scientific concepts allow for a consciousness of everyday experience, giving meaning to everyday experiences, supporting the naming and explanation of lived reality.	Scientific concepts and everyday concepts are closely connected processes that influence each other. They are two types of concepts which in the actual course of development shift back and forth many times.
3. *A cultural-historical reading of science education:* Conceptual development is the relations between everyday and scientific concepts	Everyday concepts provide the foundational real world experiences that are needed for giving meaning to scientific concepts.	Scientific concepts allow a child to be able to think and act independently of the concrete situation.	Scientific concepts and everyday concepts develop in different ways.

(continued)

Table 4.1 (continued)

	Everyday concepts	Scientific concepts	The dialectical relations between everyday and scientific concepts
4. *Science knowledge is not static:* Everyday and scientific concepts are not static they change over time and across communities.	Everyday concepts form in relation to the specific cultural communities, shaping what a child experiences and pays attention to in their everyday life.	Scientific concepts are culturally developed; they grow from particular time periods and societal needs for explanation of everyday events. There is nothing inherent in the word of a concept that gives insights into its meaning. It must be learned through the particular cultural community from which the concept arises.	Neither everyday concepts nor scientific concepts are static. What children pay attention to in their everyday lives has changed since ontogenesis. Similarly, the meanings and explanations of scientific terms have also changed, marked as scientific periods within history, such as we see in Western science of Aristotelian science, Newtonian science and contemporary science (e.g., Einstein).

Table 4.2 Affective imagination in early childhood science education

Concept	Emotionality in scientific learning
Collective scientific investigations	*Children collectively* develop a consciousness of scientific and technological *concepts and emotionality* by working together with other children to solve the problem.
	A *scientific narrative forms* as children collectively work together to solve scientific and technological problems
	In a scientific narrative, children empathise and want to help the characters to solve the collective scientific and technological problem.
Emotional imagination	Through role play of scientific narratives and learning, the *children collectively* begin to anticipate the results of each others' actions in the play, begin to anticipate their own actions, including image-bearing dramatization, verbal descriptions, prop use and transformation, and importantly, the scientific solutions created through the support of the teacher.
Flickering	Children *flicker* between concrete and imaginary worlds.
	It is the border of the imaginary world and the concrete world that creates a dialectical relation and emotional tension that promotes scientific conceptual development.
Imagination, thinking and emotions in play	Children give new meanings to objects and actions to everyday situations when learning science – creating a new scientific sense of the situation.

(continued)

Table 4.2 (continued)

Concept	Emotionality in scientific learning
Dual role of emotions in thinking	Children feeling happy enacting or exploring a science narrative with others, but also feeling excited or curious by learning new things and solving scientific and technological problems in order to scientifically help the characters in the narrative.
Emotional anticipation	In scientific investigations, children's feeling state becomes connected with the learning as they anticipate *finding a solution*. Through consciously considering feeling states in science, emotions become intellectualized, generalized, and anticipatory, while cognitive processes acquire an affective dimension, performing a special role in meaning discrimination and meaning formation (e.g., gut feeling this is going to work).
Emotional filtering	*Emotional filtering* is "where kindergarten teachers attribute emotional significance to events" (Iakovela, 2003, p. 93).
	Teachers help children in knowing what is noteworthy to pay attention to in science learning. What should they notice or look for?

References

Iakovela, E. L. (2003). Emotional mechanisms underlying personal and creative development. *Journal of Russian and East European Psychology, 41*(6), 92–100.

Vygotsky, L. S. (1987). Problems of general psychology. In R. W. Rieber, & A. S. Carton (Eds.), *The collected works of L.S. Vygotsky* (N. Minick, Trans., Vol. 1). New York: Plenum Press.

Part II
Knowledge Construction in Science

Chapter 5
Knowledge Construction in Early Childhood Science Education

Marilyn Fleer and Niklas Pramling

Abstract In this chapter we discuss forms of knowledge construction and consider these in relation to early childhood science education literature. We examine different regions of the world to see how research in science education has developed and what it has allowed us to better understand about the young learner. We consider the forms of logic drawn upon by scholars in the Australasia and South Pacific region, in the European and Nordic region, and the US. In order to understand the contributions these scholars have made for early childhood science education, we explore the forms of knowledge within the context of the child development paradigms that have underpinned science education. Our analysis of the literature draws upon a critique the following three paradigms: *the individualistic paradigm*, the *social interactionist paradigm*, and a *cultural-historical paradigm*. In this chapter empirical examples are presented in their original form from the country from which they were generated. This locates the research and theoretical concepts, but it also gives a more genuine international focus when conceptualsing science learning.

Keywords Individualistic paradigm • Social interactionist paradigm • Cultural-historical paradigm • Australasia and South Pacific region • European region • Nordic region • United States

5.1 Introduction

Carter (2007) in reviewing the vast body of empirical literature into science education makes the claim that "Despite years of formal science education, students' scientific misconceptions are common, and their lack of motivation and feelings of alienation show in the decreasing numbers opting to take science beyond the compulsory years" (p. 3). Although this is probably specific to the Western literature, and to Western constructions of science, her work does provide an interesting analysis as to what has gone wrong. She states that the "research argues the need for science education to go beyond imparting scientific conceptual knowledge and skills and advocates critical participation in a world dominated by science" that is "conceptualized by sociocultural and political interests" (p. 7). Here she suggests that science has become "mythologized" into a "scientific practice recapitulated as

M. Fleer, N. Pramling, *A Cultural-Historical Study of Children Learning Science*,
Cultural Studies of Science Education 11, DOI 10.1007/978-94-017-9370-4_5

received knowledge in school science curricula" (p. 7) where an objective and reductionist view of science is conceptualized as the *scientific method.*

Much of the literature into school science suggests that these problems can be overcome if science is more authentic for children in schools, where better learning will result when they are exposed to the "messiness of scientific knowledge construction" (Carter, 2007, p. 7). Science has become increasingly discredited as arguments about sustainability have shown, due in part to the widespread belief that science will reveal a single truth real. When multiple expert views are publicised in the media, general confusion in the community results. Carter claims that in the context of an increasingly global community, where diverse students, knowledges and practices reign, and where a fragile ecology has becoming increasingly evident, that "shared meanings have receded, and have been replaced by uncertainty and insecurity" (p. 2). Yet in this context science education has predominantly been framed as an empirical approach to knowledge construction, where it has been the tradition for "science education to be "derived from highly abstract and fragmented statements of Western canonical knowledge" (p. 2). This approach to science education foregrounds a particular form of logic, resulting in empirical knowledge that has positioned us as now being ill-equipped and unable to move forward into the twenty-first century (Carter, 2007). What this analysis highlights is the differing forms of knowledge construction possible in science.

In this chapter we discuss forms of knowledge construction and consider these in relation to early childhood science education literature. We examine different regions of the world to see how research in science education has developed and what it has allowed us to better understand about the young learner. We will consider the forms of logic drawn upon by scholars in the Australasia and South Pacific region, in the European and Nordic region, and the US. In order to understand the contributions these scholars have made to early childhood science education, we explore the forms of knowledge within the context of the child development paradigms that have underpinned science education. Our analysis of the literature draws upon a critique the following three paradigms:

1. *The individualistic paradigm*
2. *Social interactionist paradigm*
3. *Cultural-historical paradigm*

5.2 Three Paradigms for Understanding Children's Development in Science

As has been discussed in Part I of this book, a cultural-historical reading of science education in early childhood settings brings together as inseparable the individual, the environment and the social dimensions or interactions between individuals. Understanding how the child develops in this dialectical relation has been a hotly contested area within the field of early childhood education generally, due to longstanding assumptions about child development. Rather than reproduce the critiques found in the literature on approaches to child development, such as, developmentally appropriate practice, we will explore three general paradigms that underpin different forms of child development. We do this so we can better understand how knowledge construction in science learning has been framed, and why a cultural-historical reading of science education is productive for early childhood science education. But this is also important because in the broader field of early childhood education, researchers and practitioners have always engaged in critiques of theory about children's development, and historically both constructivist theories (see Chap. 1) and developmentally appropriate practices, have been left wanting.

5.2.1 The Individualistic Paradigm

Underpinning science education and early childhood education specifically has been epistemological individualism. That is, "a commitment to the notion that the mind is the outcome of processes set in motion by the individual organism" (Scribner, 1997, p. 281). The focus in early childhood education has traditionally been on finding out what the individual thinks, documenting each child's development through observations of what an individual says and does, and placing these records into individual portfolios (Fleer, 2010). This represents an *individual-world* model, where each element – individual – world – function as a natural isolated system. The world acts upon the child, and the child internalizes these actions. Scribner (1997) suggests that internalized actions "are gradually coordinated into increasingly powerful structures of thought which can be described by logical models" (p. 282). Piaget's (1972) original work demonstrated this form of logic through the example of a child placing pebbles on the ground, counting in one direction and then counting from the other direction. Wondering about the results (i.e., same number counted in both directions), the child re-arranges the pebbles and puts them into a circle, with still the same result. A level of abstraction occurs when the child is able to *deduce* from the evidence she has gathered (three different ways of counting the same pebbles) a common result, giving rise to an understanding that the sum of the elements is independent of the order of the pebbles. This elementary form of deduction laid the foundation for the child's mathematical reasoning. But this explanation is also the basis of scientific deduction. In an individualistic paradigm the social dimension is factored in, but only as retarding or accelerating the natural development of the child's deductive reasoning. The evidence used to build this theoretical approach to child development has been extensively critiqued (see Chap. 1) and is generally well known (e.g., Donaldson, 1978; Hundeide, 1985). *As will be seen when we review the literature further in this chapter, it is surprising to see that many early childhood education researchers still organise their research with an individualistic perspective in mind.*

5.2.2 Social Interactionist Paradigm

George Herbert Mead sought to interrupt the world-mind dichotomy with a sociogenetic account of how learning developed. In this logic, Mead conceptualized the development of thought as the individual *and* the world as inseparable. In this logic the development of thought was considered as occurring through the *coordination among* individuals in social interaction. For example the child who is counting pebbles is no longer conceptualized as being on her own, but rather as learning with others. The child together with other children, discuss where to start counting, from which end, or where to begin counting in the circle. If they all start counting in different places will the result be different? Scribner (1997), in citing the original experiments of Doise and Mugny (1984), states that a form of sociocognitive conflict results when different points of view are given and the children work out how to resolve the differences. *As was shown in Chap. 1, this worldview underpins the movement to socioscientific view of teaching science, where argumentation as an approach underpins secondary schools science in many countries around the world.* This was a major paradigm shift at the time because it established a new perspective in psychology and in education for a social account of cognitive development. Scribner (1997) sums up a social interactionist perspective as "cognitive development can be understood as a spiral of causality in which various cognitive preconditions in the child, which are themselves based on previous social interactions, allow the child to participate in more complex social interactions, ensuring the elaboration of more complex cognitive instructions, and so on." (pp. 283–284).

5.2.3 Cultural-Historical Paradigm

Scribner (1997) was early to recognise that social interactionist perspectives, whilst moving forward in terms of embedding children's thinking in interactions among others, that this perspective "cut off cognition from objects and actions in the world of things" (p. 283). She argued that "social interaction begats cognition which begats social interaction in ever-increasing complexity" (p. 284). This she suggested is still a bifurcated picture of child development, and both paradigms (individualism and social interactionist paradigms) "entirely ignore the larger system of social relationships and practices which constitute society and culture, and make individual transactions possible and meaningful" (p. 284). In science education, a cultural-historical paradigm offers a very different reading of how scientific learning progresses. A cultural-historical account of learning suggests that children are not simply engaging in material things, learning about their properties, but rather they are engaged in social modes of interaction where they are learning codes of behaviour and societal and family rules and activities. These are family, country and culture specific. Learning to use a spoon, your right hand, or chop sticks, to eat is not just about managing the tool, but it is also the social conventions which inform how that tool should be used within a given family, community and culture. Tool and symbol use are not learned independently of society, but rather they are part of a socially mediated process. The object or the action has no meaning without someone giving it meaning. Teachers introduce scientific tools and actions as a form of mediated action. A thermometer (object) or the scientific method (action) or a particular scientific word (sign), are all given meaning through others in socially meaningful situations in our community, family or classroom. The child's engagement in the world with others gives meaning to actions and object relative to the societal values, goals and needs. In this activity setting the child has agency and contributes to shaping how, when and where this socially produced mediation is actioned (Fleer, 2010; Hedegaard and Fleer, 2013). This is a unitary process that represents "an integrated view of human ontogeny capable of assimilating empirical findings and raising new questions" (Scribner, 1997, p. 287). Yet a close look at the literature shows that this latter view of development and learning in science has not been extensively used in early childhood science education.

We now turn to a detailed examination of the research literature on early childhood science education in order to determine how research has been framed. We draw upon the three paradigms of understanding children's development in science to analyse this literature. What do we know about the research in early childhood science education in regions such as Australasia and the Nordic region? What form of cultural knowledge has been created that we call early childhood science education? What forms of knowledge are privileged in these studies? In this chapter we seek to examine the research evidence that underpins early childhood science education across a range of countries. Specifically we review all those studies which focus on the prior to school settings, although at times we also examine science in the early years of school. We then move forward to discuss the specific nature of knowledge construction from the perspective of a more globally and culturally responsive approach for understanding early childhood children's thinking and learning in science education. The point of this chapter is to examine the forms and nature of knowledge in early childhood science education and to consider how this knowledge is constructed and privileged through the design and presentation of our research.

5.3 Australasian and South Pacific Contexts

In looking closely at what had been published about early childhood science education from 1972 onwards, we note that there are only a splattering of studies from this time period until the 1990s. Most of these fall with an individualistic paradigm where constructivism has been the dominant theory informing research. Generally, there are relatively few studies of early childhood science education in the Australasian and South Pacific region in the prior to school settings, and almost none outside of Australia and New Zealand published in English written journals.

Science education in the early childhood period in Australasia did not appear in the literature in any significant way until the beginning of the 1990s (Fleer, 2001). Even with the extensive research following constructivist approaches to science in New Zealand (e.g., Osborne & Freyberg, 1985), this work did not include the prior to school period. In 1990 Hardy and Bearlin (1990) included in their research early childhood inservice teachers, where they collectively created a gender-sensitive program for teaching science, specifically problematising the dominant empirical and traditional view of knowledge creation. How some of these teachers taught science in their preschools became the focus of research by Fleer (1990, 1991) and Fleer and Beasley (1991), but with a specific focus on children's conceptions in science in relation to how the teachers taught. The latter, although grounded in the language of alternative conceptions, drew upon cultural-historical theory to conceptualise the study and findings. These studies focused mostly on physics topics, an area that has been suggested to be outside of what most early childhood teachers were likely to teach (Fensham, 1991). In contrast Venville (2004) concentrated on topics more likely to be taught to young children, such as living things. Here the focus was on conceptual change, but from an ontological and social perspective. A social interactionist paradigm strongly influenced by cultural-historical theory begun to emerge in the science education literature. In line with the international literature at the time where alternative views held by children were considered as problematic, Venville studied conceptual change from both an ontological and a social constructivist perspective, drawing upon Vygotsky's theory of development. Venville found a number of patterns of learning relevant to conceptual change including, persistence of a nonscientific framework guiding thinking, a theoretical framework in transition, and for some, a successful radical change to a scientific framework.

In attempting to work outside of the dominant constructivist and individualistic frameworks guiding early childhood science education research, Fleer, Sukroo, and Faucett (1994, 1995) investigated Indigenous children's understandings in science, using role play and traditional stories to illicit their understandings, noting that these approaches did not specifically allow for gaining insights into children's cultural constructions of knowledge, even with Indigenous researchers guiding the study and undertaking the interviews. These studies highlighted the culturally specific nature of framing science education research. Environmental frameworks for learning science have also featured, but mostly these focus on thinking about looking after the environment, with only one specifically examining how scientific concepts aid this process (see Cutter-MacKenzie & Edwards, 2006; Edwards & Cutter-Mackenzie, 2011).

A slow movement towards a cultural-historical paradigm was emerging, but within a context of not a great deal of research into early childhood science education. For instance, in 1991 a themed issue on science and technology education was published in the *Australian Journal of Early Childhood*, representing not only the first issue on this topic, but with the exception of one paper published early in the history of the journal, no other paper on science had been published until that time. However, early childhood science education was the focus of a themed issue of *Research in Science Education* in 2003. This issue predominantly featured research from Australasian region, with Fleer and Robbins (2003a, 2003b) highlighting the shortcomings of a constructivist inspired research for investigating very young children's thinking. In their cultural-historically framed paper they argued that traditional approaches to investigating young children's thinking in science in Australia have been fraught because they privilege knowledge generation for those who use a 'question and answer' discourse. In that same issue Tytler and Peterson (2003) also discuss the limitations of previous research designs for gathering information on young children's thinking. They noted in their Australian longitudinal research that children's thinking, particularly their reasoning, is well in advance of curriculum expectations. This was also noted by Fleer (1991) in a cultural-historical study of 4-year-old children's learning of electricity, where the teacher used scaffolding techniques to support science learning. She found that "children are most receptive to learning experience which help them to understand everyday phenomena no matter how difficult the concepts are perceived to be by the adult world" (p. 102).

Robbins (2003, 2009) in drawing upon cultural-historical theory has also investigated young children's thinking in science, specifically looking at their understandings of night and day. In examining other studies in this area, where knowledge construction and interviewing approaches focused primarily upon gathering empirical knowledge following traditional approaches, less was learned about young children's thinking than when a cultural-historical approach was adopted where relational knowledge was drawn out over extended time. As Robbins (2003) states "Traditional approaches to discovering young children's ideas in science tend to isolate the individual and decontextualise thinking in order to uncover certain accepted scientific views. However, research from a sociocultural perspective recognises that cognition is a collaborative process" (p. 5). What was emerging within the very small pool of research into early childhood science education, was a concern for the social and cultural context of science, rather than a simple focus on concept formation within an individualistic paradigm for designing studies. Concerns were expressed by Segal and Cosgrove (1993) who found that more could be learned about young children's understandings of light if a broader context was used for data gathering. In drawing upon learning model of cooperative learning, informal enquiry and familiar contexts, they examined not just individual understandings of light, but sought to examine the social construction of knowledge about light and shadows. They state "Our observations of children behaving casually and even seemingly off task in groups, particularly in outside settings, belie the serious conversations occurring there" (p. 283).

The individualistic paradigm with its focus on what a child knows in science, was elaborated to include the study of the educators who worked with young children. For instance, early childhood teachers' knowledge of science was a focus for Garbett (2003), who was highly critical, stating that there is a real lack of scientific knowledge amongst early childhood teachers in New Zealand. Her study, which included teachers with cultural backgrounds of Maori, Pakeha and South Pacific Islanders, suggested that these student teachers were not aware of their lack of content knowledge. She suggested that science content knowledge is even more important for early childhood teachers to learn because the open-ended pedagogical approaches adopted require greater knowledge of conceptual understandings of science if science is to be successfully taught in those contexts. Fensham (1991) has also written extensively on the lack of content knowledge of teachers during a review of teacher education in science, specifically mentioning early childhood teachers in Australia. Fleer (2009a) in expanding on this work, but in taking cultural-historical perspective, claimed that it is not just teacher knowledge of science that is the central problem, but rather the pedagogical approach and beliefs about how children learn and develop. For instance in her case study of 4-year-old children and their teachers, she noted that the lead teacher wanted a free flow program where learning of science was to occur through the provision of materials, without direct teacher introduction or conceptual framing, and where the teacher suggested the children learn in a roundabout way. In another case study by Fleer (2009b) of 4-year-old children and their teachers, she noted that when the teachers' beliefs about teaching and learning followed a cultural-historical approach where theoretical knowledge (see Chap. 6) was being developed, that the children's learning in science was much more advanced, despite the original lack of science knowledge of the teachers about the topic being explored. That is, when the teacher actively focused on the concepts in the play-based program, both teachers and children learned more science, than if they simply organised the environment with materials (see Chap. 7). In taking a broader view of teacher knowledge, Alexander and Russo (2010) in a project known as Operation Magpie, found that teachers and children became engaged in science through investigating magpies and other birds in their environment, but their conceptual knowledge in science did not significantly improve. The social context began to feature more strongly in the study designs over time. For example, in a study that examined questions and opportunities for children's learning in science at home, as well as how the science learning in a child care centre influenced what children did at home, Fleer (1996) found that children aged 2–5 year children asked significantly more scientific questions at home than in their child care centre, despite the teaching program following an interactive approach to teaching science. The study noted that children became more curious about everyday events that could be explained scientifically, and children used the scientific language introduced in the centre in the home as a direct result of the science learning occurring in the day care centre across the topics of materials, change of state of matter, evaporation and condensation, dissolving and chemical change). These studies point out that more authentic research in early childhood education becomes possible when the research net broadens and goes beyond simply finding out teacher thinking in science, such as conceptions in biology (Edwards & Loveridge, 2011).

What we begin to see is a deeper understanding of the range of ways that the pedagogy influences thinking in early childhood science. For instance, Blake and Howitt (2012) in investigating science learning opportunities in three early learning centres noted three different contexts for learning science, as shown: "*Satisfying curiosity*, *Guided play* and *Lost opportunities* where teachers' responses about the importance of science teaching and learning varied and did not appear to match the investigations" in the two centres where science learning was happening.

Interestingly the centre where guided play was occurring this "enabled the children to advance their scientific knowledge through hands-on engagement" while the centre where children explored freely they tended to lose the "initial possibilities as children lost interest and no follow-up activities to embed the learning" were provided. In the more liberal approach to learning where the children were encouraged to satisfy their curiosity in an unstructured environment and freely use resources to "advance skills according to their own agenda . . . while being encouraged and supported by caregivers" allowed for a lot of science learning to occur (p. 297).

An individualistic paradigm has emerged in both the conceptual change literature and early childhood teacher views of child development where studies of early childhood teacher professional learning have been undertaken (Watters, Diezmann, Grieshaber, & Davis, 2001, p. 1). The results show that teachers drew upon their knowledge and beliefs of a child centred view of learning and applied this to science learning. These studies show that teachers' personal knowledge of science had increased; they gained strategies specific for teaching science; and investigations rather than experiments were found to be more useful. Importantly teachers commented on the significance of having an inservice program designed specifically for play-based settings, where teachers' confidence and competence to teach science was clearly taken into account. The study design of Watters et al. (2001) goes beyond simply documenting what teachers know in science, and reveals both personal and social factors as key to better understanding early childhood teacher knowledge of science. The specific learning needs of early childhood teachers were also considered by Howitt (2011) in her sociocultural design and piloting of early childhood science resources. An Interactive resource known as *Planting the seeds of science* was developed specifically to encourage early childhood teachers to teach science. The program was piloted across a range of early childhood centres, and the finding show that teachers were immediately engaged with the resource, stating that it filled a huge gap because the resource was designed specifically for early childhood teachers, as apposed to teachers having to adapt materials planned for non-play-based settings in primary schools (Howitt, 2011). Follow up research by Howitt, Upson, and Lewis (2011) has shown that the unit of work on forensic science in the resource represents a highly contextualized and interesting approach to teaching science in early childhood, where "providing opportunities for them to participate in scientific inquiry processes (generating questions and predictions, observing and recording data, using equipment, using observations as evidence, and representing and communicating findings) and knowledge building" (p. 54) resulted. Similarly Morris, Merritt, Fairclough, Birrell, and Howitt (2007) examined the usefulness of concept cartoons as a resource for teachers finding that them to be highly stimulating and valuable for early childhood teaching. These studies add greatly to our understandings of the special learning needs of early childhood teachers, and they recognize the unique pedagogical contexts in which these teachers work. Rather than taking a deficit view of early childhood teachers' subject content knowledge of science, these studies look more broadly at the personal and contextual factors associated with learning and teaching.

Personal and social factors have also been recognized by Hardy and Bearlin (1990) who in drawing upon an interactive approach to teaching science developed professional learning approach for both preservice and inservice teachers known as the Primary and Early Childhood Science and Technology Education Project (PECSTEP). PECSTEP was designed to improve teaching and leaning in science for both early childhood and primary teachers. The outcomes of their year long study showed that teacher interest in science and the teaching of science improved, that teacher conceptions of science and technology changed from depersonalized and decontextualised body of knowledge to becoming seen as a human endeavour, broader range of teaching strategies were employed, implicit valuing of women's experiences related to science, and changes in the personal power of the participants. Hardy and Bearlin state that "We believe that for lasting attitude change to occur there must be a change of consciousness on the part of the teacher which involves a changed understanding of the nature of scientific knowledge" (p. 150). This research recognised gender as an important factor in science teaching. Few studies have examined this area since.

In 2012 Howitt et al., in drawing upon the literature which suggested the need for specifically designed courses for early childhood preservice teacher to improve their confidence and competence to teach science, designed a course with an engineering focus where both early childhood academics and engineering academics participated in the workshops. Five principles were featured in the workshops: "acknowledgement of the place of young children as natural scientists, active involvement of children in their own learning through play and guided enquiry, recognition of the place of a sociocultural context within children's learning, emphasis on an integrated approach to children's learning experiences and the use of a variety of methods for children to demonstrate their understanding and learning" (p. 162). Pre-and post-test results of teachers' confidence to teach science improved, with a range of reasons being identified for this change. Of these teachers 82 % believed that being shown to teach science had resulted in their feeling confident to teach science, 58 % of this group also stated that knowing about resources and activities for teaching science improved their confidence, and 10 % stated that the methods of teaching science had made a difference to their confidence. In terms of preservice teachers knowledge of science concepts, survey results show that not only did they feel more knowledgeable about engineering principles, but they also better understood concepts in astronomy, energy, chemistry, and the principles of forensic science. Howitt, et al. (2012) found that "the pre-service teachers did not consider science content knowledge to be the most important reason for their increased confidence" (p. 170) where science pedagogy and the science activities were found to be more important than the science content knowledge for improving confidence to teach science. This finding is supportive of earlier research, and again points to the significance of personal and social pedagogical context of knowledge construction in the teaching of science in early childhood.

In Korea we see other priorities in early childhood science emerge. Joung (2008) in drawing upon activity theory, examined how a 5-year-old child used abductive inference in science education (logical inference to give an explanation to an unbelievable situation), where typically (intuitive experienced-based) and perceived-situations (context dependent) were considered in everyday situations at home. Inagaki and Hatano (2006) examine how 5 year old Japanese children posses a theory-like knowledge system they have termed Naïve biology, allowing them to predict and give causal reasons for biological phenomena. No other studies written in English could be found for Korea or Japan. Similarly, only one study from China on early childhood science could be found for the prior to school period (even with a search of the Chinese written literature). Liu Hui (2011) discusses not just the need for kindergarten children to learn science, but also that children should be guided "to understand the moral mission of science and promote their aesthetic experiences about science" in China (p. 66). Rather than a focus on only conceptual knowledge, Liu Hui suggests that science knowledge must be learned within a moral framework, and the aesthetic dimensions should also be foregrounded. This orientation is missing from the Australian and New Zealand studies, which make up the bulk of the literature found across the Asia pacific region.

In the studies reviewed for this region, we note that knowledge construction has been conceptualized by many following a cultural-historical tradition and methods for gathering data were noted to be broader than traditional approaches in science education research during that period, with some actively problematising the nature of Western and male constructions of science. Insights from China show that more attention should be paid to the moral and aesthetic dimensions of science, something that is absent from all of the literature reviewed. Mostly the studies reviewed for the region have tended to focus on either what children thought about specific science topics, or how confident their teachers were to teach science, and how they may be supported through both preservice and inservice programs or specifically designed resources. What we do learn from the review of those studies available is that Indigenous knowledges were rarely examined. Of significance is that most researchers made references to empirical science content knowledge, but problematised the nature of this knowledge construction for early childhood education or used cultural-historical theory to conceptualise their work, where the development of theoretical knowledge is foregrounded. We now turn to the Nordic context in order to explore the research into science learning of early childhood children in that region so that we can see what forms of knowledge construction exist. We specifically examine how knowledge construction is shaped by the paradigms in which the studies are framed and undertaken.

5.4 Nordic Research on Early Childhood Science Education: A Cultural-Historical Paradigm

In this section we will review research studies on children's science learning from the Nordic countries, that is, Denmark, Finland, Iceland, Norway, and Sweden. In the Nordic countries, children tend to spend a great deal of their time at their preschool outside (see Einarsdóttir & Wagner, 2006, for texts on early childhood education in the Nordic countries; also Moser & Martinsen, 2010). It is common to make excursions to the forest or play outside even in the wintertime. Hence, there are ample opportunities for preschool teachers and children to explore and converse about nature and natural processes. Still, there is little research on young children's science learning. There are very few studies of direct relevance to the present volume. In this brief review, we will mention studies we have found but will only discuss, in some length, a limited number of studies, which are of particular interest to the themes of the present volume. Other studies (such as Thulin, 2010; Thulin & Pramling, 2009; Pramling, 2010) will be referred to more extensively in the following chapters, in relation to our discussion.

The majority of studies on young children's science learning during the last years are based on a theoretical frame that is, more or less, in line with the theory of the present book, cultural-historical theory. However, in the Nordic countries, when studying children's science learning, this perspective is commonly referred to as a socio-cultural perspective. Perspectives that are somewhat adjacent to this perspective (see e.g., Greeno, Collins, & Resnick, 1996, for an overview), such as a pragmatist perspective (e.g., Jakobson & Wickman, 2007, 2008; Klaar & Öhman, 2012) and the multimodal perspective of Gunther Kress and his colleagues (Kress, Jewitt, Ogborn, & Tsatsarelis, 2001) are also used (e.g., Elm, 2008; Elm Fristorp, 2012), if to a lesser extent. One reason for the dominance of socio-culturally informed studies may be the widely read and influential book, *Lärande i praktiken: Ett sociokulturellt perspektiv* [English: Learning in Practice: A Sociocultural Perspective] as published in Swedish by Roger Säljö in 2000. The book, which can also be read by people in Norway and Denmark as well as by many people on Iceland and in Finland has subsequently also been translated to several Nordic languages. The book was pivotal in introducing what for many was then a new perspective into educational research in the Nordic countries, including research on science education. The perspective that has dominated science education for a long time with older children and adolescents (STCSE database), that is, cognitive and/or developmental psychology is not prominent in the research on younger children's science learning in the Nordic countries.

5.4.1 Communicative, Contextual and Institutional Embeddedness

Sträng and Åberg-Bengtsson (2009) studied a group of 5-year-old children together with their teacher visiting a science centre. There they attended an exhibition called 'Way of the water', consisting of a large-scale model that you walk through, that follows the flow of water "from the mountain range in the uplands of northern Sweden down to the Baltic Sea through a number of environmental and cultural settings abstracted from the Swedish landscape" (p. 14). The children were followed attending the exhibition during the guidance of a guide from the center and later back at their preschool at circle-time when they discussed with their teacher what they had experienced at the centre. The researchers pose three questions: First, what content was focused on; second, what communicative strategies the guide and teacher, respectively, used when talking with the children about the exhibition; and third, the different ways that the interaction between the adults and the children were contextually framed.

The children participating in the study worked with their teacher for a prolonged time with the theme 'water' in various ways, for example visiting a brook nearby their preschool. During their visit to the science centre, they were accompanied by their teacher, a science centre guide and four parents. The parents and the teacher each wore an audio-recording device, recording talk between children and between the guide, teacher and children. The teacher also took photographs during their walk along the model. The researchers describe the model in the following way:

> The children entered the exhibition through a dark corridor, where the sound of thunder was heard, before climbing a staircase to the highest point of the model, where they met a Sámian teepee in front of a relief of a mountain with some (plastic) snow on the top. No water ran from this point of the model, but there was a brook painted on the relief. The sounds of rippling water as well as howling wolves and singing birds were heard. At the bottom of one flight of stairs, there was a pool with fish typically found in streams in the northern parts of Sweden. Still further down the 'Way of the water' was a beaver's lodge. In some places running water could be seen, while in others it could only be heard. (ibid., p. 19)

(The Sámi, whose teepee is referred to, are the indigenous people of northern Europe: Sweden, Norway, Finland, and parts of Russia.) After the visit to the science centre, as already mentioned, the teacher had a follow-up discussion with the children at circle-time a week later. Analysing these two learning situations (the guided tour at the science centre and the follow-up discussion at the preschool), Sträng and Åberg-Bengtsson found three different communicative patterns (i.e., ways of interacting) that they suggest are related to different contextual framings. These three patterns are illustrated and analysed in the article. But briefly described, the first pattern consists of "providing facts" and was used by the guide at the science centre; the second pattern identified, "directing attention by posing questions" were used by the teacher at the science centre; and the third pattern, "asking for accounts" was used by the teacher at circle-time. An example of the first pattern was that the guide told the children that "It's called a glacier. It is snow up there..." (p. 21). An example of the second pattern was the teacher asking the children "what's that?" and "what do you see?" (p. 22), while looking at a bird (a great crested grebe) in a pool. And an example of the third pattern was the teacher asking the children "is there anyone who remembers where the water went then?" (p. 25). In addition, while the teacher repeatedly asked the children about how the water ran, "there are no instances in our data where the children express the idea of the larger scale, coherent model. On the contrary, they talked only about individual parts of the exhibition" (p. 26). However, this may not be unexpected, since, as the researchers point out, "neither the guide nor the teacher tried to explain the model of the 'Way of the water' or scaffold the children's making of meaning of the flow of water in a more elaborated manner" (p. 28). There may be several reasons for this observation, as the researchers reason, including the model being taken for granted and therefore being left for the children themselves to 'discover'. The idea that the child, him or her self, should discover principles of nature and science (e.g., understand the model of the system) is common in discussions about children's science learning, as informed by an individualistic paradigm. However, to expect the children to discover this rather complex model (cf. the description of it above, clarifying that, for example, the water cannot be seen all the way), including understanding that "the pool with the great crested grebe represented the sea (or the Baltic Sea to be more precise)" (p. 27) is not realistic. The importance of teachers scaffolding children's sense-making through 'pointing out and linguistically informing their experiences' (Pramling & Pramling Samuelsson, 2011) is implied. This is a theme we will return to throughout the present book. Finally, Sträng and Åberg-Bengtsson (2009) conclude that children's development of 'model thinking', that is, in our alternative terms, managing representations of various kinds, is a field in much need of research into young children's science learning. This is a theme we will investigate in this book (see particularly, Chap. 10).

5.4.2 *Tool-Mediated Inquiry into the Natural World*

Ärlemalm-Hagsér, E (2008) conducted a study in order to provide developmental opportunities for and follow the development of children's understanding of insects. A cyclical design was used for the study. In brief, the approach meant to (i) try to investigate the children's experiences of insects at their preschool ground, (ii) create a learning situation where the children get to draw and talk about insects, and (iii), create a second learning situation, on the basis of the outcomes of the first one, in the forest where the children study and draw insects. The preschool teachers asked the children, "What insects do we have on the preschool ground? Draw some of these" (p. 72, our translation). Looking at the children's drawings and listening to them talking about these, it became clear that the children had a rich view of insects/"small creeping things", including ladybirds, beetles, earwigs, ants, bumblebees, shield bugs, spiders, woodlice and earthworms. They thus showed a wide awareness of different animals. The children also showed a good insight into the animals' anatomy, as evident in their drawings. Most children drew the animals from a birds-eye view. However, as seen in the list of animals depicted, the children did not differentiate between insects and other small animals. Hence, some of the challenges now facing the preschool teachers were how to support children in discerning insects as a particular species, and how to make children draw the animals (also) from a different perspective than the birds-eye view, in order to make visible other parts of the animal (on children's drawings in science education, see Chap. 10). The teachers were also self-critical about their own knowledge of the domain and how they could communicate more productively with the children, asking better questions, to further challenge the children's thinking.

Building upon what was found in the first step of the study, the teachers then tried to provide a new learning situation, introduced in the following way:

Teacher: I know that you know many things here. Yes, a gread deal and I'm sure you're wondering many things, and today I was gonna ask you this: What are insects? What is that? Anybody knows?
Nils: They're small creeps.
Maria: They're this small (showing with her thumb and forefinger, appr. one millimeter). (p. 74, our translation)

According to the researcher, the children show the sizes of insects with their fingers and hands, resulting in a span from approximately 1 mm to 8 cm. The children are then asked to draw the insects.

Teacher: We thought that we'd like to see when you draw your insects. How do the insects look?
Nils: You mean super super super enhanced? (p. 75, our translation)

Nils' question about magnification is highly relevant to the task, but the teacher does not answer him. Continuing talking with each other and the teacher, the following takes place after a while:

Nils: Guess what insect it is?
Teacher: Someone who lives in a hill perhaps?
Maria: An ant.
Nils: Ants haven't two legs, but it has four legs.
Maria: Spider.
Nils: Spiders have eight legs.
Teacher: Is it an insect that has two legs?
Nils: No, four.
Teacher: An insect that has four legs?
Nils: One, one, one foot less than I'm years [old], when I'm five.
Maria: Ladybird.
Nils: Right (to Maria),'cause these here were dots (points at the lines on the upper part of the drawn body), those were the eyes and there were the legs (showing). Everything was super duper enhanced, if it should've been them, super duper enhanced. (p. 75, our translation)

In this excerpt, the issue of the number of legs of different animals is introduced. However, nothing more is made of this relevant feature at this time. Instead, other features important to an evolving understanding of insects and other animals come to the fore in the talk:

Teacher: What is that on yours (directed to Nils, who has drawn a ladybird)?
Nils: That's the eyes, there was a nose before but I erased it.
Teacher: Why did you erase it then?
Nils: 'Cause ladybirds don't have any nose.
Teacher: No, that's right, insects don't have noses.
Nils: I haven't drawn any mouth.
Teacher: No, do they have any mouths then?
Nils: I don't think so.
Teacher: You don't think so. How can they survive?

Nils: Snails don't have any mouth.
Teacher: Snails don't have any mouth?
Maria: Yes, I've seen that.
Nils: No-o.
Maria: They have eyes anyway. (p. 76, our translation)

However, this topic is at this time not further followed up in the teacher-child talk. In reviewing the learning situations afterwards, in additions to the conclusions drawn from the initial mapping of the children's knowledge of insects (and other small animals), the issues of the number of legs of different animals (anatomy) as well as their living conditions (e.g., food) is decided to be given more consideration by the teachers on the subsequent occasion. The teachers also decide to introduce a categorization key, which consists of pictures and text that makes it possible to identify different species of animals (similar keys exist for deciding plants and mushrooms). During the third time they talk about insects, the children and their teachers go into the adjacent forest and look for insects. Using their categorization key, they are able to investigate under loupe the animals they find. The children are greatly enthusiastic about the possibility of analyzing the animals in terms of their number of legs and whether they have or have not got wings. Together they try to see whether the animals they find have six legs and three body parts. The children once more make drawings of the animals they have found and compare these to the categorization key.

Looking at these new drawings and listening to what the children have to say, it becomes clear, Ärlemalm-Hagsér, E (2008) suggests, that the children have developed their knowledge of animals, including anatomy and variation among animals. Nils, who previously (see above) did not think that insects had a mouth, now draws an ant while exclaiming, "But how many legs has an ant, six, 1, 2, 3, 4, 5, 6 (counting the legs he has drawn)… I drew my ant with 1, 2, 3 body parts!" (p. 79, our translation). In addition to having discerned the number of legs and body parts, his drawing depicting an ant now also has a mouth.

Concluding the study, the children's knowledge of insects and other animals can be described in terms of an increased differentiation, being able to differentiate out numbers of legs, body parts, and other features. This small-scale study also illustrates how talking about what one does (e.g., while making a drawing of an insect) provides developmental opportunities, not only between teacher and child but also between children (cf. above, the example whether insects have mouths). Working in a cyclical way, that is, following up on children's uptake and ideas on subsequent activities means that learning is not reduced to one-offs. Supporting children in making connections between these events, that is contextualizing backwards and forwards (cf. Mercer, 1995) is important in making sure the children make such connections and see how things relate to one another. Not only identifying what children know, about, in this case, insects, or provide opportunities for children to interact with each other, but also introducing mediating tools, such as the categorization key into a meaningful situation in which children engage, could be seen as an activity underpinned by a cultural-historical paradigm.

5.4.3 The (Missing) Practices of Early Childhood Science Education

In her study, Elm (2008, cf. Elm Fristorp, 2012) investigates how a natural science topic is selected and orchestrated in a preschool and a preschool class (an intermediate form of schooling between preschool and school for the 6-year-olds). A preschool group (one teacher and six children aged 2–4) and one preschool class group (two teachers and 14 children aged 5–7) were followed with a video camera when working on various natural (scientific) phenomena. In the study, data and analysis from four different activities are presented: floating and sinking, and "small creeping things under rocks" (in the preschool) and ants (Camponotus) and black woodpeckers" and a stuffed green woodpecker (in the preschool class). The emerging activities are analysed in terms of language use (speech) and natural science activities. Analysing the kinds of scientific activities the children and teachers engage in in their interaction, Elm points out that some basic acts such as 'planning', 'interpreting' and 'explaining' are missing from her data. For example, as she writes concerning an activity in preschool where it was tested whether different objects float or sank, "There is no reasoning about why objects float or sink" (p. 52, our translation). To large extent, "predicting tends to be left out when the children observe and examine whether objects float or sink" (loc. cit.). In our alternative terms, what appears missing from the activities studied by Elm is talking about what lies beyond (i.e., is more general than) the present instance (cf. Chap. 5), that is, how to explain what happened (retrospective speech) and how to anticipate what may happen (prospective speech).

The theme "Small creeping things under rocks" consists of children and their teacher making an excursion to a nearby forest. One of the activities they engage in is looking at insects under rocks. One thing that is evident in the empirical excerpts is that the teacher often responds to children's questions by posing a new question (e.g., 'What do you think?') or suggesting that they could 'investigate', take a look, rather than giving an answer to what the child asked in a more strict sense. This was also observed in Thulin's (2010) study, where she suggests that this may be an indicator of teachers in preschool nurturing an ideal of children finding out about the world through exploring it (see further, Chap. 2; cf. also our discussion in Chaps. 1 and 7 on the difference between a Piagetian and a Vygotskian perspective on learning and development, see also, Fleer, 2009a). This stance, in terms of the three paradigms introduced above, is underpinned by epistemological individualism.

One of the activities followed by Elm (2008), as we have already mentioned, was ants (Camponotus) and black woodpeckers. Having walked to a nearby forest, the children and their teachers among other things investigate a fallen tree trunk and a hollow stump surrounded by wood splinters. The teacher has brought along a book for interpreting traces of different animals. Together the teacher and the children compare the pictures in the book with the stump they look at to try to find out what may have made the wood splinters. The teacher suggests that this may have been caused by black woodpeckers trying to get to ants (Camponotus) in the stump. Elm writes:

> More elaborate explanations and reasoning about what one does and why are missing. Further activities of an investigative nature that activates the children are also missing. The teacher's comments appear to be spontaneous responses to needs at the moment. When a child answers a question, the teacher follow up the child's answer with another question that is often introduced by, Do you think… There is no sustained reasoning of an overarching kind, for example, about why the children think it can be the way they express. In the discussions that occur, the teacher appears to be making an inventory of the children's ideas. Such discussions are concluded when a child delivers the answer the teacher expects. (p. 75)

While being rather concerned with what was 'missing' from these early childhood science education practices, Elm's study does say something about what such activities consist of for the children participating, and thus what is made possible for them to learn. What learning opportunities different educational practices offer children is important to investigate with an interest in children's science learning. Rather different learning opportunities will be seen in other early childhood science education activities that we investigate in other chapters in this book.

Research into early childhood science education in the Nordic countries is much in line with the perspective taken in this book, in investigating early childhood science activities from a socio-cultural (cultural-historical) perspective with a particular focus on communication and other tool use. There is thus an affinity between this research and the research from Australia informed by a cultural-historical perspective that we reviewed in the previous section. In the next section we will look at empirical studies of, and discussion pieces on, early childhood science education from Greece and the US.

5.5 Greece: A Social Interactionist Paradigm

One of the few countries from where it comes quite a few studies on early childhood science education is Greece. In this section we will therefore review some recent and fairly recent studies that are of interest to the present book.

In his discussion of early childhood science education, Hadzigeorgiou (2001) argues that in order to establish a foundation for the child's science learning, "certain attitudes do facilitate its establishment and it would be preferable to start with helping young children develop these attitudes" (p. 64). He goes on to argue that certain attitudes towards science "are the prerequisites or the motivators for children's engagement in science activities" (p. 64). What he refers to as attitudes particularly concerns "intellectual curiosity" (p. 64). Hadzigeoriou's reasoning is made against the tradition of 'pedagogically appropriate' activities, and while he states that he recognizes this approach as sound, he suggest that it may not establish any long-term relationship between the child and science. According to this reasoning, there may be activities that are not 'pedagogically appropriate' that should still be included and emphasized in early childhood science education; these are activities that "can make children feel perplexity, wonder, amazement and surprise without the possibility of their direct action on objects and subsequent investigations" (p. 65). Some of the activities intended to incite wonder into children are: "Emptying water from one glass into another using a piece of towel cloth without moving or tilting the glasses"; "Inflating a balloon by putting it on the top of a bottle that is left for a while in the sun"; and "Making an egg float on the surface of water by putting more and more salt in the glass" (p. 65). These kinds of activities are suggested to provide "great stimulus for learning" (p. 65), especially for preschoolers. In his reasoning, Hadzigeorgiou refers to empirical observations where such activities have been conducted, but the paper contains no information on what was observed and how in more systematic terms. Hence, this paper should be considered a discussion piece, rather than an empirical investigation into early childhood science education. Referring his reasoning back to philosopher of science Alfred North Whitehead (1861–1947), who, Hadzigeorgiou writes,

> believed that in order for students to be able to reflect on knowledge that will not be inert, a certain rhythm of its presentation should be followed by teachers. To describe the rhythm he used the terms 'romance', 'precision' and 'generalisation'. Children should begin their engagement with any subject in a 'romantic' way, i.e. in a way that makes them feel the excitement inherent in the subject. (p. 66; cf. Rule, 2007)

Asking himself how such 'romance' could be induced, Hadzigeorgiou suggests through stories. It is further suggested that through, for example, a story about the tension between 'hot' and 'cold', the child will learn the concept of 'cool' (p. 67). However, from our point of view, it does not seem clear how "binary opposites" such as "energy as something good and energy as something bad" (p. 67) would develop children's understanding of energy as a science concept. And building on the works of Bruner (1990, 2006; as also, albeit briefly mentioned by Hadzigeoriou), the differences and complex relationship between a narrative account and a paradigmatic (scientific) one needs much more theoretical elaboration and empirical study. Still, with these comments, the importance of nurturing children's interest in the phenomena of nature (as explained by science) should not be underemphasized; we will return to this issue from a different point of view in Chap. 11.

How to initiate preschool children to science is also the topic of another paper, by Ravanis and Bagakis (1998). The problems of this paper are how an appropriate curriculum for preschool could be developed and what teaching strategies should be used. Contrasting what is referred to as an "empiricist" perspective (a kind of object-manipulation and instruction approach) with a sociocognitive (Doise & Mugny, 1984) one, the authors argue the merits of the latter; such as the importance of social interaction and negotiation between partners. More specifically, this approach is said to hold merit over the alternative due to the communication between the children "leads to the decentration from the subjective perspective" and children "facing the arguments of a collaborator understand that for a question of a problem there are many possible solutions, consideration and strategies of dealing with it" (p. 319). Hence, a social interactionsist paradigm underpins the discussion. Following their reasoning about different approaches to early childhood science education, Ravanis and Bagakis illustrate a teaching sequence concerning the gasification of water.

Exploring preschool pedagogic practices related to science, Tsatsaroni, Ravanis, and Falaga (2003) use sociology of education theorist, Basil Bernstein's work to argue and illuminate that "the emergent discourse of pre-school teaching and learning of specialized content is in tension with dominant pre-school pedagogic practices, and that the contradictory demands placed upon teachers" to focus more on science content, on the one hand and to provide a play-based activity on the other "might lead to a narrowing of the view of learning in pre-school classrooms" (p. 385). This perspective is used to discuss a pilot study conducted in nursery school on magnetic properties and materials susceptible and not susceptible to magnetic attraction. Tsatsaroni et al. (2003) argues that this emerging tension between discourses place contradictory demands on the teacher:

> Thus, pre-school teachers might shift between a pedagogy that constructs weak boundaries between specialized school knowledge and everyday knowledge, based on the ideological notions of play and activity as a means of developing the child, and characterized by slow pacing, invisible criteria and interpersonal forms of control; to one which constructs strong boundaries, puts an emphasis on 'lesson' as specialized content, and is characterized by strong pacing, and too narrow criteria of evaluation of the practice (and pupils). (p. 412 f.)

If so, this would fundamentally rearrange the nature of early childhood (science) education.

In an experimental study, Ravanis, Christidou, and Hatzinikita (2013) investigated children's understanding of light. Two groups of in total 170 preschool children (approximately 6 years old) were studied, with pretest, teaching intervention and post-tests. One group of children participated in activities built on the principles of a sociocognitive approach, while the other group participated in activities on the basis of what is referred to as "an empiricist perspective" (p. 1). In the sociocognitive group, "a familiar metaphor was introduced in order to facilitate children to construct a 'precursor model' about light" (p. 1); the metaphor being "the travel of light through space" (p. 9). The distinction made between the two approaches is explained as "The empiricist approach is based on the conviction that the provision of organized stimuli (activities) to children can ensure learning while the sociocognitive approach attempts to support children in constructing a precursor model based on the use of a familiar metaphor" (p. 4). The findings indicate that both groups of children developed their understanding of light from pre- to post-tests but that the "cognitive progress" (p. 1) made by the children in the sociocognitive group was more significant than the progress made by the children in the empiricist group. "These results," Ravanis et al. (2013) suggest, "indicate the significant contribution of the teaching activities involving interactions that were structured around the existing obstacles to children's cognitive development" (p. 17). Hence, this study serves to emphasize the important role of others, including the teacher, not only for organizing the environment for children but also for reasoning with children.

5.6 The United States of America: An Individualistic Paradigm

From an American point of view, Baldwin, Adams, and Kelly (2009) suggest that many early childhood teachers “are struggling with the notion of how to blend an instructional focus on academic content standards with the National Association for the Education of Young Children’s (NAEYC) 12 Principles of Learning and Teaching that have been identified as preferred practice for the field” (p. 71 f.). Against this background, Baldwin et al. describe “an approach used by one university supported demonstration school to develop an assessment supported, child-centered, and emergent curriculum framework that addresses both preschool content standards and developmental domains” (p. 72). Central to “emergent curriculum”, they suggest, is “maintaining a commitment to build instruction on children’s interest” (p. 72). While the importance to build upon children’s interest and sense-making is important; we would argue that institutions such as preschool and school are also a society’s way of ensuring that children are introduced to fields of knowledge and develop new interests than they would have in their home environment. Furthermore, taking a Vygotskian perspective, typically institutional forms of knowledge such as ‘scientific concepts’ (Vygotsky, 1987) build on other principles – abstract systems made up by relations between concepts – than what the child has experiences of. This does not mean that the child’s previous experiences are not important to appropriating such institutional forms of knowing, but the relationship is complex (we return to this issue throughout this book). About the demonstration school, Baldwin et al. (2009) write: “Based on a sound understanding of child development and learning, the team determined that the most engaging and therefore most efficient way for young children to learn is when instruction builds upon their interest. Staff believed that topics of learning are best garnered through the ideas, excitement, and questions of the children themselves” (p. 72). This reasoning is underpinned by an individualistic paradigm, but, as we have already hinted at, the possibilities of building on children’s interest and/or also having to interest children to new forms of knowing and phenomena is a complex issue that needs to be considered. In addition, as we will discuss in some length in Chap. 11, asking questions is in itself something that children develop through participation in an activity – or a prolonged theme/project – rather than necessarily having beforehand. “Children’s natural curiosity with the world around them and the questions they ask often related to science concepts”, Baldwin et al. suggest about the children in the demonstration school they write about. To encourage exploration in the children, naturally occurring events, for instance, finding worms after it has rained, were used as starting points. What is referred to as a ‘science concept planner’ is then constructed by the teachers, starting with the science concept, its related concepts and materials and standards relating to these concepts are identified. This approach, they write, “differs from the common practice of choosing activities and then determining what can be learned from them” (p. 74). During children's exploration, their progress is documented by the teachers and displayed on a project board and in children's individual portfolios.

In a statistical study, based on longitudinal data with over 8000 children, Saçkes, Trundle, Bell, and O'Connell (2011) investigated the impacts of selected early science experiences in kindergarten on children's achievements at the beginning and end of kindergarten and in third grade. The availability of science materials was found to facilitate teaching of science and children's participation in such activities. "Children's engagement with science activities that involved using science equipment", Saçkes et al. report, "was not a significant predictor of their end of kindergarten science achievement. However, children's participation in cooking activities was" (p. 217). Summarising their results, Saçkes et al. suggest that their study indicates that "early childhood experiences provided in kindergarten are not strong predictors of children's immediate and later science achievement" and that this limited effect may be due to "the limited time and nature of science instruction" (p. 217) in kindergarten. What is in the study referred to as "science materials" are exemplified with "water and sand table, and science or nature area with manipulatives" (p. 220). However, these materials and environments do not say anything about how they were used or engaged with, and therefore, whether science activities evolved, we would argue. For example, concerning "children's science activities" with "science equipment (e.g., magnifying glass, scales, thermometers) and cooking and food related items" (p. 222), the following is stated:

> Using a magnifying glass to examine insects or rock samples, measuring quantities and temperature, and using food related items to develop measuring skills and to study properties of matter are typical science activities in kindergarten classrooms, and these activities involve scientific skills. Therefore, these variables were used as the indicators of children's science activities in the study. (p. 222)

A study of this kind raises many questions. Whether it is sound to measure children's science scores in kindergarten must surely be questioned. Furthermore, the issue of testing children's understanding is a very complex issue (see e.g., Chap. 7 of this book for a discussion). It must also be questioned if more science teaching (as estimated by the teachers) is necessarily better, that is, more developmental for children than the nature of such teaching. For a forceful argument to the contrary, that it is how teachers and children communicate in science activities that is decisive, see Fleer (1995; see also, Gustavsson & Pramling, 2014). In fact, Saçkes et al. (2011) themselves suggest that "future studies also should examine the nature of teacher – child interaction in science learning in early years" (p. 229). "Children do not learn science in early years because few science learning opportunities are provided for them" (p. 230) with the majority of the teachers of Saçkes et al.'s (2011) study report that they teach science once or twice a week.

In a discussion piece, Brenneman and Louro (2008) argue that what they call 'science journals' can be used in preschool for supporting and assessing children's science and literacy learning (cf. Chang, 2012). They also discuss the importance of teachers talking with the children about their journal entries (primarily drawings), suggesting that "[f]rom a Vygotskian perspective, teachers model the sorts of questions children may ask themselves as they record observations, providing a scaffold for children's learning" (p. 115). Furthermore, representing observations in journals may motivate children to observe attentively, providing an incentive for doing so.

5.7 Conclusion

As seen in the review we have made in this chapter, early childhood science education is today a concern for research and scholarly debate. What was shown was that there were different perspectives on how such education should be organized to provide for children's learning, and research in this field is informed by more or less distinct theoretical traditions – individualistic paradigm; social interactionist paradigm; and cultural-historical paradigm

It can be argued that scientific knowledge can be conceptualized as being located within specific areas of everyday life, such as school knowledge, work-based knowledge, or knowledge about how to do things at home. In this conception, knowledge encompasses practices where problems arise and solutions need to be found, where goals are met, and new possibilities created. As suggested by Hedegaard and Chaiklin (2005) "General knowledge refers to that knowledge which is used commonly to address these [problems, goals, possibilities] needs" (p. 52). General forms of knowledge that are created in one arena, and which have evolved over time and used in another arena, are forms of *societal knowledge* (Hedegaard and Chaiklin).

In the science education literature reviewed, knowledge construction is about a specific form of societal knowledge often named as subject-matter content or academic knowledge or discipline knowledge. But these forms of knowledge cannot be considered as independent of what Hedegaard and Chaiklin (2005) have called *local knowledge*. In their conception, local knowledge is a kind of knowledge that is created at home and in the community. This was evident in the studies reviewed in this chapter on early childhood teacher knowledge of science, where specific localized ways of teaching and learning featured as an important finding – such as play based programs.

What also featured in this chapter was the significance of personal knowledge of the early childhood teachers. Personal knowledge is similar to Vygotsky's (1987) theory of everyday concepts and science content knowledge as related to Vygotsky's theory of scientific concepts (see Chap. 1). How teachers and children related to the science content knowledge and turned this into personal knowledge, was featured as important in the Australian studies, but few concentrated upon how to solve this problem. This line of enquiry was more evident in the Nordic countries, for example, in terms of how to respond to children's questions (see also further our empirically-based discussion about the latter matter in Chap. 11). While the Nordic studies generally were underpinned by what we refer to as a cultural-historical paradigm, findings from early years practices illustrated how teachers may base their pedagogy on an individualistic paradigm. This line of inquiry was non-existant in the studies undertaken in Greece, and only slightly touched on in the US, where an individualistic paradigm for framing research was prevalent.

Both societal and personal knowledge interacts with subject matter knowledge (Hedegaard & Chaiklin, 2005). The nature of this interaction should allow subject matter knowledge to become personal knowledge for the child to use in everyday life, and not just in school contexts. According to Hedegaard and Chaiklin (2005) "How children's personal knowledge from home and community life will be related to academic knowledge in school depends on the form of academic knowledge and the teaching practice" (pp. 52–53). In the next chapter we take up this challenge and examine teaching pedagogy in early childhood settings where narrative, empirical and theoretical knowledge emerge as a result of the practices and beliefs of the early childhood teachers.

References

Aikenhead, G., & Michell, H. (2011). *Bridging cultures: Indigenous and scientific ways of knowing nature*. Don Mills, Canada: Pearson.

Alexander, A., & Russo, S. (2010). Let's start in our backyard: Children's engagement in science through the natural environment. *Teaching Science, 56*(2), 47–54.

Ärlemalm-Hagsér, E. (2008). Insikter om insekter – naturvetenskap i förskolan [Insights into insects: Natural science in preschool]. In I. Pramling Samuelsson & N. Pramling (Eds.), *Didaktiska studier från förskola och skola* [Didactic studies from preschool and school] (pp. 67–81). Malmö, Sweden: Gleerups.

Baldwin, J. L., Adams, S. M., & Kelly, M. K. (2009). Science at the center: An emergent, standards-based, child-centered framework for early learners. *Early Childhood Education Journal, 37*, 71–77.

Blake, E., & Howitt, C. (2012). Science in early learning centres: Satisfying curiosity, guided play, or lost opportunities. In K. C. D. Tam, M. Kim, & S. W. Hwang (Eds.), *Moving forward: Issues and challenges in science education research* (pp. 281–299). Dordrecht, The Netherlands: Springer.

Brenneman, K., & Louro, I. F. (2008). Science journals in the preschool classroom. *Early Childhood Education Journal, 36*, 113–119.

Bruner, J. S. (1990). *Acts of meaning*. Cambridge, MA: Harvard University Press.

Bruner, J. S. (2006). Narrative and paradigmatic modes of thought. In *In search of pedagogy, Volume II: The selected works of Jerome S. Bruner* (pp. 116–128). London: Routledge.

Carter, L. (2007). Sociocultural influences on science education: Innovation for contemporary times. *Science Education,* 1–17. doi:10.1002/sce.

Chang, N. (2012). The role of drawing in young children's construction of science concepts. *Early Childhood Education Journal, 40*, 187–193.

Coulson, R. (1992). Development of an instrument for measuring attitudes of early childhood education towards science. *Research in Science Education, 22*, 101–105.

Cutter-MacKenzie, A., & Edwards, S. (2006). Everyday environmental education experiences: The role of content in early childhood education. *Australasian Journal of Early Childhood, 22*(2), 13–19.

Doise, W., & Mugny, G. (1984). *The social development of intellect*. Oxford, UK: Pergamon.

Donaldson, M. (1978). *Children's minds*. Glasgow, Scotland: Fontana/Collins.

Edwards, K., & Loveridge, J. (2011). The inside story: Looking into early childhood teachers' support of children's scientific learning. *Australasian Journal of Early Childhood, 36*(2), 28–35.

Edwards, S., & Cutter-Mackenzie, A. (2011). Environmentalising early childhood education curriculum through pedagogies of play. *Australasian Journal of Early Childhood, 36*(1), 51–9.

Einarsdóttir, J., & Wagner, J. T. (Eds.). (2006). *Nordic childhoods and early education: Philosophy, research, policy, and practice in Denmark, Finland, Iceland, Norway, and Sweden*. Greenwich, CT: Information Age.

Elm, A. (2008). *Interaktion och naturvetenskap i en förskola och en förskoleklass* [Interaction and natural science in a preschool and a preschool class] (Licentiate thesis). Stockholm: Institutionen för didaktik och pedagogiskt arbete.

Elm Fristorp, A. (2012). *Design för lärande – barns meningsskapande i naturvetenskap* [Design for learning: Children's meaning-making in science]. Stockholm: Stockholm University.

Fensham, P. (1991). Science education in early childhood education: A diagnosis of a chronic illness. *Australian Journal of Early Childhood, 8*(3), 3–11.

Fleer, M. (1990). Scaffolding conceptual change in early childhood. *Research in Science Education, 20*, 114–123.

Fleer, M. (1991). Fusing the boundaries between home and child care to support children's scientific learning. *Research in Science Education, 26*(2), 143–154.

Fleer, M. (1995). The importance of conceptually focused teacher-child interaction in early childhood science learning. *International Journal of Science Education, 17*(3), 325–342.

Fleer, M. (1996). Socially constructed learning in early childhood science education. *Research in Science Education, 21*, 96–103.

Fleer, M. (2001). Science over 40 years. *Australian Journal of Early Childhood, 26*(1), 39–45.

Fleer, M. (2009a). Supporting scientific conceptual consciousness or learning in 'a roundabout way' in play-based contexts. *International Journal of Science Education, 31*(8), 1069–1089.

Fleer, M. (2009b). Understanding the dialectical relations between everyday concepts and scientific concepts within play-based program. *Research in Science Education, 39*, 281–306.

Fleer, M. (2010). *Early learning and development: Cultural-historical concepts in play*. New York: Cambridge University Press.

Fleer, M., & Beasley, W. (1991). A study of conceptual development in early childhood. *Research in Science Education, 2*, 104–112.

Fleer, M., & Robbins, J. (2003a). Understanding our youngest scientific and technological thinkers: International developments in early childhood science education, Editorial. *Research in Science Education, 33*, 399–404.

Fleer, M., & Robbins, J. (2003b). "Hit and run research" with "hit and miss" results in early childhood science education. *Research in Science Education, 33*, 405–432.

Fleer, M., Sukroo, J., & Faucett, T. (1994). Determining young Aboriginal children's scientific understandings: A pilot study. *Research in Science Education, 24*, 371–372.

Fleer, M., Sukroo, J., & Faucett, T. (1995). A cross-cultural study of 4–8 year old Aboriginal children's scientific understandings: A pilot study. *Australian Research in Early Childhood Education, 2*, 73–81.

Garbett, D. (2003). Science education in early childhood teacher education: Putting forward a case to enhance student teachers' confidence and competence. *Research in Science Education, 33*(4), 467–482.

Greeno, J. G., Collins, A. M., & Resnick, L. B. (1996). Cognition and learning. In D. C. Berliner & R. C. Calfee (Eds.), *Handbook of educational psychology* (pp. 15–46). New York: Macmillan.

Gustavsson, L., & Pramling, N. (2014). The educational nature of different ways teachers communicate with children about natural phenomena. *International Journal of Early Years Education, 22*(1), 59–72.

Hadzigeorgiou, Y. (2001). The role of wonder and 'romance' in early childhood science education. *International Journal of Early Years Education, 9*(1), 63–69.

Hardy, T., & Bearlin, M. (1990). Outcomes of the primary and early childhood science and technology education project at the University of Canberra. *Research in Science Education, 20*, 142–151.

Hedegaard, M., & Chaiklin, S. (2005). *Radical-local teaching and learning a cultural-historical approach*. Aarhus, Denmark: Aarhus University Press.

Hedegaard, M., & Fleer, M. (2013). *Leaning, play and children's development: How families and schools shape children's learning*. New York: Cambridge University Press.
Howitt, C. (2011). Planting the seeds of science. Development and evaluation of a new flexible and adaptable early childhood science resource. *Teaching Science, 57*(3), 32–39.
Howitt, C., Upson, E., & Lewis, S. (2011). 'It's a mystery!' A case study of implementing forensic science in preschool as scientific inquiry. *Australasian Journal of Early Childhood, 36*(3), 45–55.
Howitt, C., Blake, E., Calais, M., Carnellor, Y., Frid, S., Lewis, S. W., et al. (2012). Increasing accessibility to science in early childhood teacher education through collaboration between teacher educators an science/engineering academics. In K. C. D. Tam, M. Kim, & S. W. Hwang (Eds.), *Moving forward: Issues and challenges in science education research* (pp. 157–173). Dordrecht, The Netherlands: Springer.
Hui, L. (2011). On the children science education in the kindergarten. *Studies in Preschool Education, 5*, 44–66.
Hundeide, K. (1985). The tacit background of children's judgments. In J. V. Wertsch (Ed.), *Culture, communication, and cognition: Vygotskian perspectives* (pp. 306–322). New York: Cambridge University Press.
Inagaki, K., & Hatano, G. (2006). Young children's conception of the biological world. *Current Direction in Psychological Science, 15*(4), 177–181.
Jakobson, B., & Wickman, P.-O. (2007). Transformation through language use: Children's spontaneous metaphors in elementary school science. *Science & Education, 16*, 267–289.
Jakobson, B., & Wickman, P.-O. (2008). The roles of aesthetic experience in elementary school science. *Research in Science Education, 38*, 45–65.
Joung, Y. J. (2008). A 5-year-old's initiation into scientific ways of knowing. In Y.-J. Lee & A.-L. Tan (Eds.), *Science education at the nexus of theory and practice* (pp. 317–332). Rotterdam, The Netherlands: Sense Publishers.
Klaar, S., & Öhman, J. (2012). Action with friction: A transactional approach to toddlers' physical meaning making of natural phenomena and processes in preschool. *European Early Childhood Education Research Journal, 20*(3), 439–454.
Kress, G., Jewitt, C., Ogborn, J., & Tsatsarelis, C. (2001). *Multimodal teaching and learning: The rhetorics of the science classroom*. London: Continuum.
Mawson, L. (2007). Introduction: Bridging the cognitive and sociocultural approaches in research on conceptual change: Is it feasible? *Educational Psychologist, 42*(1), 1–7.
Mercer, N. (1995). *The guided construction of knowledge: Talk amongst teachers and learners*. Clevedon, UK: Multilingual Matters.
Morris, M., Merritt, M., Fairclough, S., Birrell, N., & Howitt, C. (2007). Trialling concept cartoons in early childhood teaching and learning of science. *Teaching Science, 53*(2), 42–45.
Moser, T., & Martinsen, M. T. (2010). The outdoor environment in Norwegian kindergartens as pedagogical space for toddlers' play, learning and development. *European Early Childhood Education Research Journal, 18*(4), 457–471.
Osborne, R., & Freyberg, P. (1985). *Learning in science: The implications of children's science*. Auckland, New Zealand: Heinemann.
Piaget, J. (1972). Development and learning. In R. E. Ripple & V. N. Rockcastle (Eds.), *Piaget rediscovered*. Ithaca, NY: Cornell Press.
Pramling, N. (2010). Unearthing metaphors: Figurativeness in teachers-child talk about soil and related matters. *Early Childhood Education Journal, 38*(1), 57–64.
Pramling, N., & Pramling Samuelsson, I. (Eds.). (2011). *Educational encounters: Nordic studies in early childhood didactics*. Dordrecht, The Netherlands: Springer.
Ravanis, K., & Bagakis, G. (1998). Science education in kindergarten: Sociocognitive perspective. *International Journal of Early Years Education, 6*(3), 315–327.
Ravanis, K., Christidou, V., & Hatzinikita, V. (2013). Enhancing conceptual change in preschool children's representations of light: A sociocognitive approach. *Research in Science Education*. doi:10.1007/s11165-013-9356-z.

Robbins, J. (2003). The more he looked inside the more Piglet wasn't there: What adopting a sociocultural perspective can help us see. *Australian Journal of Early Childhood, 28*(2), 1–7.

Robbins, J. (2009). Analysing young children's thinking about natural phenomena: A sociocultural/cultural historical perspective. *Review of Science, Mathematics and ICT Education, 3*(1), 75–97.

Rule, A. C. (2007). A "tad" of science appreciation. *Early Childhood Education Journal, 34*(5), 297–300.

Saçkes, M., Trundle, K. C., Bell, R. L., & O'Connell, A. A. (2011). The influence of early science experience in kindergarten on children's immediate and later science achievement: Evidence from the early childhood longitudinal study. *Journal of Research in Science Teaching, 48*(2), 217–235.

Scribner, S. (1997). Three developmental paradigms. In E. Tobach, R. J. Falmagne, M. Brown Parlee, L. M. W. Martin, & A. Scribner Kapelman (Eds.), *Mind and social practice: Selected writings of Sylvia Scribner* (pp. 281–288). New York: Cambridge University Press.

Segal, G., & Cosgrove, M. (1993). "The sun is sleeping now": Early learning about light and shadows. *Research in Science Education, 23*, 276–285.

STCSE database (http://www.ipn.uni-kiel.de/aktuell/stcse/stcse.html).

Sträng, M. H., & Åberg-Bengtsson, L. (2009). 'From the mountain and then?' Five-year-olds visiting the 'Way of the water' exhibition at a science centre. *International Journal of Early Childhood, 41*(19), 13–31.

Thulin, S. (2010). Barns frågor under en naturvetenskaplig aktivitet i förskolan [Children's questions during a science activity in preschool]. *Nordisk Barnehageforskning, 3*(1), 27–40.

Thulin, S., & Pramling, N. (2009). Anthropomorphically speaking: On communication between teachers and children in early childhood biology education. *International Journal of Early Years Education, 17*(2), 137–150.

Tsatsaroni, A., Ravanis, K., & Falaga, A. (2003). Studying the recontextualisation of science in pre-school classrooms: Drawing on Bernstein's insights into teaching and learning practices. *International Journal of Science and Mathematics Education, 1*, 385–417.

Tytler, R., & Peterson, S. (2003). Tracing young children's scientific reasoning. *Research in Science Education, 33*(4), 433–466.

Venville, G. (2004). Young children learning about living things: A case study of conceptual change from ontological and social perspectives. *Journal of Research in Science Teaching, 41*(5), 449–480.

Vygotsky, L. S. (1987). *The collected works of L. S. Vygotsky, Volume 1: Problems of general psychology, including the volume* Thinking and Speech (R. W. Rieber & A. S. Carton, Eds., & N. Minick, Trans.). New York: Plenum.

Watters, J. J., Diezmann, C. M., Grieshaber, S. J., & Davis, J. M. (2001). Enhancing science education for young children: A contemporary initiative. *Australian Journal of Early Childhood, 26*(2), 1–7.

Chapter 6
Knowledge Construction Is Culturally Situated: The Human Invention of Empirical, Narrative and Theoretical Knowledges

Marilyn Fleer

Abstract In this chapter the different forms of knowledge construction that are typically illustrated in the literature are examined alongside of data that have been used to illustrate what they look like in practice. Three forms of knowledge are introduced: narrative, empirical and theoretical. Paradigmatic thinking and dialectical thinking are discussed in the context of generating scientific knowledge. Examples from both the science education literature and a study of preschool children learning about mixing materials are given. However, knowledge construction in these forms is not a common framework for reporting (or even discussing) reports in science education research. As such, studies which demonstrate different forms of knowledge construction in science learning are drawn upon and used alongside of empirical data generated by the preschool children studying the mixing of substances (empirical and narrative) and the form and structure of insects (theoretical) found in the outdoor area in the preschool. The latter highlights both commonplace practices found in preschools for teaching science, and discusses the challenge of introducing empirical knowledge in a play-based setting and puts forward evidence on how theoretical knowledge can be introduced to young children.

Keywords Knowledge construction • Empirical knowledge • Narrative knowledge • Theoretical knowledge

6.1 Introduction

> The word *science* was deliberately chosen to replace *natural philosophy* during the political birth of a new organization in 1831: the British Association for the Advancement of Science … As a result of the evolution of natural philosophy into professional science, present-day science is strongly based on Euro-American thinking. . . most scientists' professional culture is Eurocentre in character, and can be described as *Eurocentric science* or *Western science* (Aikenhead & Michell, 2011, pp. 21–22).

Aikenhead and Michell (2011) suggest that there are many forms of knowledge that are culturally specific, and that Western science is one form of knowledge. So what kinds of knowledges do we privilege in early childhood settings when we

M. Fleer, N. Pramling, *A Cultural-Historical Study of Children Learning Science*, Cultural Studies of Science Education 11, DOI 10.1007/978-94-017-9370-4_6

teach science? This chapter examines three forms of knowledge that children can and do develop through the study of science in the early childhood settings. We begin by specifically introducing case examples of how empirical and narrative knowledge are formed in early childhood centres. We then contrast this with the development of theoretical knowledge for children through presenting a case example of science teaching in one centre were the focus was on paradigmatic thinking. The point we wish to make in this chapter is that most early childhood teachers privilege narrative thinking and learning when they do not take an active role in science education, leaving it to the resources 'to do the teaching of science'.

What do we mean by empirical, narrative and theoretical knowledge? These forms of knowledge construction are introduced in the next section through case examples to illustrate these knowledge forms in relation to the pedagogy used for teaching science to young children.

6.2 Empirical Knowledge

Empirical knowledge and paradigmatic thinking has been discussed in relation to science education (as well as other subject matter areas), through the metaphor of building blocks. Blocks of knowledge are learned in school science or discovered in the scientific community, and these blocks of knowledge build one on top of the other. These blocks of knowledge are abstracted concepts. According to this knowledge tradition, they are formed as a result of close observation, descriptions of those observations are made, classification of what has been observed is undertaken, and some form of quantification to document what has been discovered results. The assumptions underpinning empirical knowledge is that knowledge can be observed, quantified, presented as an accurate representation of what was observed, and understood as abstracted concepts, and then used away from the site of the original observation. The building up of this knowledge over time, like the blocks in a tower, continues unless one of the blocks or information is proven wrong.

In science curricula these building blocks of knowledge are ones that students must learn if they are to acquire the necessary science knowledge deemed important within both the scientific and education community within a particular society. Blocks of knowledge are often categorised around specific content areas within science, such as biology, physics, and chemistry. How these content areas in science relate to each other may not always be the focus of attention, but rather knowing the science knowledge (building block) is what dominates in many schools, and this has been a source of criticism, blamed for turning students off learning in the sciences. Hedegaard and Chaiklin (2005) have suggested that "If instruction is based only on empirical knowledge it will orient pupils to acquiring concepts from different subject domains that are not related to each other or to their local life world" (p. 54).

The disassociation of knowledge from the site of its construction, as is how science content knowledge is commonly conceptualized, is also a problem for teachers. For example, early childhood teachers who learn concepts in isolation from their construction, as blocks of knowledge to acquire, also find it difficult to then

Table 6.1 Interview of teacher beliefs about constructing empirical knowledge

Teacher	….parent's will find out that their children are learning more than just numbers and that….outside they didn't call it potions and I actually heard them use the word stuff…...**I'd rather the children didn't say this is a potion….. they didn't have fixed word for it….**
Research Assistant	The potion play went on too……it all flowed from one thing to the next and the next… from cooking to poisoning to siphoning…
Teacher	…it all just evolved….
Research Assistant	**…..so the potion could be anything…it's a non specific word……generic…and assumes that transformations can happen.**
	The leaves today went ….to cooking, perfumes, and experimenting with water, smell….various sequences…..but scientific words I didn't hear much
Teacher	There are children coming out and in …when they want… I really liked the independence…I **did not set** up one thing ….the children did it all themselves…..and **I was really pleased** with **that because I just think people set things up too much for the children.**

Adapted from Fleer (2009a)

work out how these knowledges can be taught to young children. In the following interview about a science teaching program on materials, we see that the preschool teacher was keen to introduce science activities to her children, but she found it difficult herself to know what were the concepts and how the concepts could be reproduced in the activity, or how knowing the concept could solve a problem or address a personal need of the children. Rather, her focus was on just setting up the environment to see what might happen, what the children would do with the materials. In the example, we see that the teacher and the assistant teacher gave a different perspective on how to organise science learning for the children (Table 6.1).

Without adult suggestions about what to do with the materials, the teacher believed that the resources themselves would generate learning opportunities.

In the actual teaching program the teacher provided oil, water, vinegar, and shaving cream for the children to mix. The teacher also placed an array of pumps, buckets, different sized containers, water, and dyes for the children to explore, as shown in Fig. 6.1. This activity was named by the children as 'potions'. She wished the children to learn about mixing substances together, as an activity to support science learning. But the teacher did not frame the experience in any particular way. Rather she simply provided the materials, as a form of discovery learning.

This approach to teaching and learning in early childhood education is commonplace. Teachers generally do not set up controlled experiments for the generation of empirical knowledge. The teacher's interactions with children in these situations is about supporting the children's free exploration of the materials, perhaps drawing their attention to what is happening as they are mixing the materials. The experiences remain

Fig. 6.1 Children explore materials by mixing substances together

at the everyday level, because no system for focusing the children's attention on the materials occurs, no descriptions of those observations are made via photographs or drawings, the classification of what has been observed is not undertaken or discussed across the group, and no form of quantification to document what has been discovered results. Consequently, no empirical knowledge is formed for the children, but rather a deepening of everyday concepts of these everyday materials results.

In the example that follows, we observe how the teacher introduced above through the interview, explored materials in the sandpit with a group of children (see Fleer, 2009b). The teacher placed a range of items on the edge of the sandpit for the children to use in their free play. The children had oil, vinegar, shaving cream, water, and sand, plus a range of containers. In the first part of this transcript, the teacher labeled Lana's play as an experiment. Lana also used this language.

6.2.1 *Observation: Mixing Oil in Sandpit (26.8)*

Teacher: "*Um it's it's Lana's oil experiment*" (Lana pouring oil into a bowl, puts oil down, makes sure the lids on and then turns the oil container so that the label is facing her).

Lana: "*There*". (picks up oil container and looks at the label)

"*Baking, oil experiment*" (Puts oil container down and picks up container with mixture and walks to another area where Molly is playing. Lana puts sand into her container and swishes it around. Molly then gets up and goes to where Lana was. The child then brings the oil back and starts pouring it into another container).

"Oh this is working babe" (Lana looks up as she speaks, walking back to a pretend oven and puts ingredients into the space).

The teacher sat alongside of the two children – Molly and Lana – and interacted with them as they poured or sprinkled into the different containers the materials available to them. Molly took an oil bottle over to her teacher and asked her to close the lid.

Teacher:	*"Ah that is hard to shut isn't it?"* (Molly tries to push lid down).
Molly:	*"Hard to shut".*
Teacher:	*"So what are you going to do with it?".*
Molly:	*"Shake it".*

The teacher begins to direct the children's attention to the materials in the containers. The teacher's focus of attention was on 'mixing' substances and using their senses to notice any changes. However, as becomes evident in the interactions of the child with the teacher, Lana's focus of attention was on cooking meat. She had created an imaginary situation of cooking in the sandpit. The oil and vinegar containers and their actual contents, suggested 'cooking' to these children – cooking meat specifically.

Lana:	(Child is mixing ingredients in a bowl). *"I'm going to mix this* (Teacher: *hm-hm*) *all the way to the bottom, to the end."*
Teacher:	*"What does it smell like?".*
Lana:	*"Um, cause I'm making meat".*
Teacher:	*"You're making?".*
Lana:	*"Meat".*
Teacher:	*"Meat okay"* (Lana stops mixing and pours oil in). *"More oil?".*

The teacher tried to explicitly point out to Molly and Lana that the substances were not mixing together. Molly attributed this to the physical difficultly of mixing. Molly's focus of attention returned to cooking, and this time she suggested that she was 'making different kinds of oil'.

Teacher:	*"What can you see Molly what can you see?"* (tilts container).
Molly:	*"Oh water and oil".*
Teacher:	"What's this at the top?" (Molly looking). *"Can you see how something's at the top and there's other stuff at the bottom and then?".*
Molly:	*"There's oil* (points to top) *there's um water* (points in middle) (Research Assistant-yep) and there's sand (points to bottom)".
Teacher:	*"Why do you think it does that?".*
Molly:	*"Cause I put it in there I put them all in there".*

Teacher:	*"Yeah but why do they all stay layered I thought you shook it?"* (Molly starts shaking).
Molly:	"I couldn't shake it properly".
Teacher:	*"You can't shake it properly well how about we shake it together* (shaking together) *here we go. We're doing really well together aren't we?".*
Lana:	*"Yeah we make some more different oil".*
Teacher:	"Okay let's have a look at it".
Lana:	"I make some more different oil".
Teacher:	"See we shook that didn't we Molly but it's still the same".
Lana:	"I make some more" (comes over to Molly and teacher and observes).
Molly:	"Yeah but it ?" (pushes on lid).
Lana:	"I make some different oil".
Teacher:	"Okay you made some different oil" (Lana pours oil into bowl and Molly looks on).

The teacher worked hard to re-direct the children's attention from making meat to looking at the mixing of the oil, water and sand. The children took note, but focused on 'making different oils'. The activity did not support scientific thinking, but rather provided the children with a playful event where they expanded their experiences of playing with cooking oil. Later the teacher asked the children to comment again on the materials in the mixture, but the response from the children indicated that they had reframed the experience in relation to cooking once more:

Lana:	"Put a little bit, a little bit more sand (grabs a handful of sand and puts it into bowl) little bit, mix it all around".(Picks up handful of sand with other hand and puts it into the bowl) "Lots of sand" (Mixes then picks up oil and pours it into bowl).
Teacher:	A different type of oil.
Lana:	(Puts oil down and grabs something else and puts it down next to the oil. Opens up oil and stands up).
Teacher:	"How come there's all these spots in it?" (Pointing in bowl, Lana leans forward and looks into bowl).
Lana:	"Oh cause that's my meat" (Stands up and walks away with oil).

The children used their everyday concepts of these substances provided, having seen them used in cooking, in order to contextualise their experimentation. Their investigations in these playful events focused on mixing but not in relation to developing a scientific understanding about the nature of materials, but rather through pretending to be cooking meat. Because the children were in an imaginary situation, they were not thinking about the resources in relation to the concept of materials and their properties they were using to support their cooking. The teaching program was not organized to build empirical knowledge through a systematic approach to knowledge construction using a form of scientific method suitable for preschool children. Rather the children were left to make sense of the materials on their own, and when the teacher joined the children it was not possible for them to leave their imaginary situation and to focus on the real attributes of the materials.

This example highlights both commonplace practices and the challenge of introducing empirical knowledge in a play-based setting. It also shows why it is important for a teacher to be clear about what empirical knowledge s/he wishes to introduce. That is, if the focus is on exploring the materials (and not setting them up in a particular scientific way) then it is important that the experiences will allow for a particular kind of scientific concept to be discovered. In the example given, the teacher introduced a range of materials, but the combination of these materials through the children's mixing of them, did not necessarily lend themselves to generating empirical knowledge through discovery learning.

Whilst empirical knowledge is highly valued in society, as demonstrated by the fact that most countries have a science education curriculum of some kind, there are limitations to this form of knowledge construction, as evidenced by the way this knowledge is taught in preschools and schools (Carter, 2007). We now turn to another form of knowledge construction that is common in preschools – narrative knowledge.

6.3 Narrative Knowledge

Jerome Bruner in his book *Actual minds, possible worlds*, conceptualised an epistemology for a proposed set of characteristics of narrative knowledge and thinking. He argued that "We know the world in different ways, from different stances, and each of the ways in which we know it produces different structures or representations, or, indeed, "realities." (p. 109). Bruner argued back in 1983 that "Narrative deals with the vicissitudes of human intentions" and stories contain well-formed realizations (p. 16) of these vicissitudes. With narratives, arguments that are pro and cons are deemed more interesting than conclusive. Knowledge construction in this form is about constructing a convincing story. As Bruner (1986) states that:

> In the *telling* there must be "triggers" that release responses in the reader's mind, that transform a banal fibula into a masterpiece of literary narrative... Whatever the medium-whether words, cinema, abstract animation, theater-one can always distinguish between the fibula or basic story stuff, the events to be related in the narrative, and the "plot" or just, the story as told by linking the events together (p. 19).

What we see emerging in narratives is a dual landscape, where both reality and fantasy occur concurrently as human plight is contemplated in the narrative:

> the reader is helped to enter the life and mind of the protagonists: their consciousness are the magnets for empathy. The matching of "inner" vision and "outer" reality is, moreover, a classic human plight (pp. 20–21).

Narratives have their own internal structure and logic for building characters. For instance:

> . . . in the folktale, character is a *function* of a highly constrained plot, the chief role of character being to lay out a plot role as hero, false hero, helper, villain, and so on. For while

it may be the case that in the time-smoothed folktale story-stuff determines character (and therefore character cannot be central), it is equally true that in the "modern" novel plot is derived form the working out of character in a particular setting (on of the earliest theorists of modernism, therefore, being Aristotle on tragedy!)" (p. 20).

Structures such as plight is also significant:

the fibula of story-its timeless underlying theme-seem to be a unity that incorporates at least three constituents. It contains a *plight* into which *characters* have fallen as a result to intentions that have gone awry either because of circumstances, of the "character of character," or most likely of the interaction between the two. And it requires an uneven distribution of underlying consciousness among the characters with respect to plight. What gives the story its unit is the manner in which plight, characters, and consciousness interact to yield a structure that has a start, a development, and a "sense of an ending". Whether it is sufficient to characterize this unified structure as *stead state, breach, crisis, redress* is difficult to know. It is certainly not *necessary* to do so, for what one seeks in story structure is precisely how plight, character, and consciousness are integrated (p. 21).

Narrative dialogical thinking helps children to conceptualise experience and construct personal meaning that can be transcended from situated experience to general human and societal life. To do this, Bruner (1986) worked with three categories for formulating a narrative method. For instance, *presupposition* captures the idea of creating implicit meaning. Implicitness dominates, rather than explicit meaning. *Subjectification* is foregrounded in the narrative method, where reality is constructed through personal subjective narratives rather than objective processes. Multiple perspectives are also valued. Instead of a single universal truth, narratives feature different perspectives expressing segments or parts of a constructed reality. Bruner argued that "we become increasingly adept at seeing the same set of events from *multiple* perspectives or stances" (p. 109). These 'folk theories' of everyday events are built and expressed through a range of media, and this represents knowledge construction and models of thinking of daily life that is common among most young children. What children gain is an internal form of logic that is principled.

Bruner (1986) in contrasting narrative knowledge with empirical knowledge construction, states that with the formulation of the latter through experiments, the knowledge generated tells "us nothing about the discourse that converts an unworded narrative into powerful and haunting stories" (p. 19). Most experiences of the world go beyond documenting events and actions into rational or scientific knowledge, where accounts must be "replicable, interpersonally amenable to calibration and easy correction" (p. 110). An example of this form of principled knowledge construction in the same preschool described above is discussed further below (Fleer, 2009a, 2009b). In the example that follows, the children did not make meaning of the objects as intended by the teacher. Rather than producing empirical knowledge in science about an array of scientific concepts, the children made meaning of the situation by drawing upon a known narrative of a nursery rhyme of humpty dumpty as a form of narrative knowledge, extending it further to include an activity of medicating humpty dumpty as a way of 'repairing humpty after falling off the wall'.

6.3.1 *Transcript: Medicine for Humpty Dumpty (23.8)*

Three girls at a table outside, they have two plastic bottles one has a spoon in it the other has a pump action dispenser. There is a Humpty Dumpty soft toy nearby.

Jayde:	He fell off the wall again and this is a girl Humpty
Lana:	Humpty fell off the wall again
Grey Girl:	Wait I'll spray it I have to spray it. (takes spoon out and puts it under the dispenser and fills spoon)
Jayde:	Oh hi ah Humpty Dumpty
Lana:	Hello
Grey Girl:	Here you go (passes spoon to Jayde)
Jayde:	Hello how are you today (Another child wearing cream jacket joins)
Cream Girl:	Ah let me see. (Comes over to table, is holding a mobile phone in one hand and touches Humpty Dumpty's arm) touch it here.
Green Girl:	Yes he's dead, he's dead I knew he he's dead. (climbs onto table, little girl with black jumper leaves).

The children had to draw upon known narratives in order to make personal meaning of the materials provided by the teacher, because the materials, made no sense to the children, and the teacher did not introduce a conceptual framework to the children for generating empirical knowledge. Rather what happened was that the group of children used the narrative of Humpty Dumpty to bring their everyday understandings of medicine together with their understandings of healing Humpty Dumpty who has fallen off the wall. Potions for these children was not about materials and their properties to be gleaned through mixing, but rather it was about medicine and caring for people in the community. The conceptual focus for the children was personal, and their way of working with the materials and knowledge construction was narrative.

As often happens in preschool settings, the science activities provided by the teacher were used by the children in ways unintended by the teacher, as the children explored common personal experiences in their play of being given foul tasting medicine. Many teachers acknowledge that this will happen, often stating "Let's see what they will do with this" or "Let's find out where these materials will lead the children". Narrative knowledge construction and thinking is common in early childhood settings, and fits with the pedagogy of a play-based curriculum, where a great deal of role-play occurs. The example above not only illustrates how children create narrative knowledge in early childhood settings, but it demonstrates how personal knowledge construction in early childhood science education occurs when adults are not involved, or minimally involved in the process.

In the next section we examine another form of knowledge construction that includes empirical and narrative knowledge construction, but draws upon a different form of logic for realizing science learning for young children. The case example examines theoretical knowledge and paradigmatic thinking.

6.4 Theoretical Knowledge

We begin this section by introducing Davydov's (2008) theorisation of the term 'concept' followed by a discussion of how he used this term in theoretical knowledge construction with children. According to Davydov a concept can both *represent* a material object *and* be used to *reflect* on that material object. The concept allows for a particular mental action to occur. To do this, a child must first be aware of the material object, in order to form a conscious mental representation of that object. For example, the air that surrounds children will be experienced intuitively as part of transpiration and as a force when they are running or riding their bikes or playing with prams and toy cars. However, children will not necessarily consciously consider the air that surrounds them, let alone factor it into their play as one force that is acting upon their toys. We know from research that young children do not consciously consider air, or even contemplate air as a material (see Sere, 1985).

It is only when children consciously consider an object, that they can give it a new meaning. In preschool settings children are already well practiced in giving new meaning to objects in their play, such as when a stick becomes a hobby-horse or when a box becomes a car (Fleer, 2011). In science education, teachers also want children to give new meanings to objects in their environment, but they wish for children to develop a scientific meaning of that environment. In play-based programs what is needed is teacher mediation to frame or draw attention to the natural environment as affording scientific meaning (see Chap. 2). As we noted above, teacher mediation is critical for helping children to develop a scientific meaning of their environment.

Davydov (2008) argues that concept formation that successfully builds theoretical knowledge and dialectical thinking is about a system of concepts that are relationally linked and relationally understood. For example, young children regularly interact with their natural environment when playing in the outdoor preschool area, but are unlikely to consciously realize that they are a part of a natural ecosystem. Children find things in their environment, both at preschool and at home. They will look under leaf litter, sheets of tin, stones and logs and discover all kinds of insects. They may observe these insects, re-discovering these kinds of insects in other contexts. To build theoretical knowledge requires a particular kind of mediation by the teacher, so that children look with scientific eyes, as they build an understanding of their finds in relation to the ecosystem. Research by Fleer (2011) has shown that to achieve this, the teacher needs to:

1. determine what might be the core scientific concepts to be learned;
2. engage children in considering both the particular (e.g., ant), and the general (species classification)
3. support children to re-create their learning as models (often rudimentary)
4. *rise to the concrete* by having the opportunity to consider how the abstract knowledge (e.g., species classification) was formed in the first place (observing form, function, food source, and habitat of a particular insect)

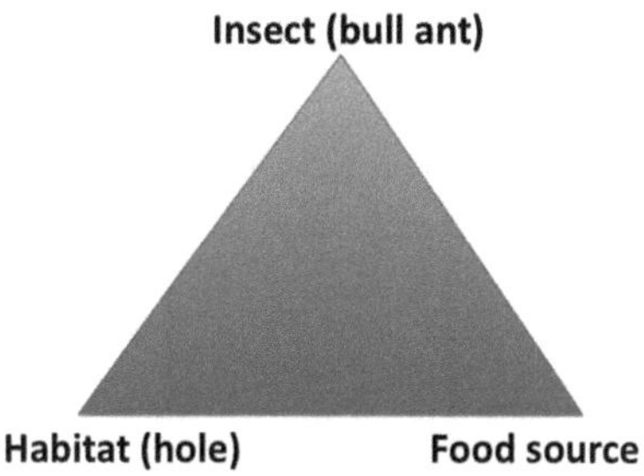

Fig. 6.2 Determining core concepts – beginning with the child's personal interest of a "bull-ant in the wrong place"

Core Concepts Davydov (2008) argues that for theoretical knowledge construction to occur, that the essence of the concept must be determined. What really matters for concept formation when a child finds an ant in the 'wrong place'? What concept or theoretical knowledge could the teacher develop in this situation? Unlike the example of the teacher who provided materials for mixing substances, where the children used and developed narrative knowledge, the following example (Fleer, 2010, 2011) illustrates theoretical knowledge construction. The teacher did not just provide materials to the children to see what would happen, rather in the following example the teacher specifically considered the essence of the scientific concept she was seeking to develop. The teacher considered the child's comment about the bull-ant being in the wrong place, and used this as an opportunity to build theoretical knowledge about an ecosystem, where relational understandings is central.

The teacher considers the child's find (i.e., the ant) and determines what might be the core concepts for building theoretical knowledge and dialectical thinking. That is, she considers an ecosystem where habitat, structure of the insect, and food are all related (Fig. 6.2). The teacher determines the core concepts within a system of concepts that s/he believes are necessary for the child to build relational knowledge. For instance, looking at the relations between what the child finds, the habitat in which it was found, and the food sources available. This rudimentary ecosystem is a theoretical model that helps children move beyond single and disconnected forays when exploring their environment to a more systematic conceptual investigation of their natural environment.

With theoretical knowledge of an ecosystem, children explore their environment in a particular way. Davydov (1990) in drawing upon Davydova has argued that theoretical knowledge 'always pertains to a *system of interaction,* the realm of successively connected phenomena that, in their totality, make up an organized whole' (p. 254).

In the field notes that follow (see Fleer, 2011 for details of the study), we describe a context where a child has found an insect in 'the wrong place'. The teacher used the opportunity to introduce investigative tools, such as magnifying glasses, insect boxes, and binoculars, to frame how children engaged with their environment. The teacher conceptualised the experiences that follow, by supporting the idea of a map, and the task of mapping the finds.

6.4.1 Map and Treasure Hunt

Christian adapts a treasure hunt activity from the day before and takes the map he's made and marked with an X inviting Teacher J to follow him outside to hunt for bugs.

Teacher J: *Should we go and find the path?*
Christian: *Yes…*

Christian has spent time each day looking carefully around the yard with binoculars and magnifiers but today he is the trying to use the abstracted view of the yard that his map represents to locate bug treasure at point X. This is a new experience and challenge and he seeks support from his teacher to embark on this venture.

Christian: *…(can we find it) …without the map*
Gale: *I gave something to Christian.* (Gail hands Christian something to encourage his treasure hunt search in the environment)

All four children follow Christian and the teacher (14.2).

Davydov (2008) explains that children's investigations begin as 'flashing impressions', where elements of significance are singled out or are conceptualized as the 'essence of the thing' being observed. That is, children may notice that a specific insect can be found in specific locations within the preschool, such as a slater under rotting wooden logs, or ants coming out of ant holes. Knowing about the *relational link* between insect and habitat as a rudimentary model for an ecosystem, valued by Western science, is an important concept for children to learn.

In building theoretical knowledge, what is to be developed is not just an understanding about a particular insect, but rather a concept of an insect, within a system of relational concepts (insect form and structure, habitat, food source) which together make up the universal concept of 'an ecosystem' and 'classification system of living things', as is detailed in many science curriculum documents.

Dialectical Relations Between the Particular and the General Children need to consider both the particular ant, and the general conceptualization of insects in an ecosystem. Building theoretical knowledge is also about the particular and the general. Davydov (2008) stated that children need practice at concurrently thinking about the particular (individual organism – e.g., ant), and thinking about the general system of concepts (e.g., insect as a classification system). Investigating an outdoor area of a preschool, creating a map, allows children to move from the general to the particular, and from the particular to the general – as a dialectical process. Here children also concurrently deal with the imaginary situation of the map and the real situation of the outdoor area. In the case example introduced above, the teacher also used books and photocopied sheets of insect classifications to support Christian's investigation of insects in the preschool environment. The field notes show how the teacher moved Christian from the particular ant to a more general conception of insects.

6.5 Naming Bugs (27.2)

Teacher J has charts and insect identity sheets as resources for children in the centre who want to name the bugs they find. Christian has found a 'bug' and believes it to be a centipede. He brings it indoors for clarification of identification. Christian looks closely at the chart and points to and names, the Centipede, Mosquito, Praying Mantis and Lacewing.

Christian:	I think that's a centipede
Teacher J:	I think that's a centipede. Yep. I'll read the word centipede yep that one's a centipede. That one's a millipede. They're the ones we find around the kinder all the time.
Colleen:	We found one. Sticks on. I think it will go through those holes.
Christian:	Mosquito.
Teacher J:	That one's called a scorpion fly.
Christian:	Praying mantis.
Teacher J:	Special names.
Colleen:	Praying mantis.
Teacher J:	Yep.
Christian:	Lacewing.

The naming of small creatures represents a bringing together of aspects of children's scientific knowledge (as Christian shares his understandings) and observational knowledge of the insects the children have actively sought, uncovered, cared for and played with, in their environment. Davydov (2008) argued that the essence of the learning must also be crystallised into a model. That is, examining resources without actively constructing a model of the essence of what is being investigated would not go far enough in the quest for developing theoretical knowledge.

Modeling Representing thinking as a model is possible within play-based programs because resources and time are readily available for engaging in drawing, painting, collage and box construction. In the example of the bull-ant in the wrong place, the teacher invited the children to represent their understandings as action drawings, paintings and collage. In reproducing the form and function of the ant in relation to it's habitat, Christian created a 'pac-man munching machine' and a 'bull-ant going to the dentist'. Although not fully functioning models, these examples illustrate how children make meaning and document their growing understandings of relational concepts as a rudimentary model. The field notes and transcripts of modeling making are shown in Table 6.2.

Rising to the Concrete Modeling helps children to *rise to the concrete*. Rising to the concrete encompasses the pedagogical principle of initially examining a holistic system and mentally ascending to this system in order to determine its specific nature. Through establishing the individual relations it is possible to observe its universal character. Through this kind of contemplation, children discover a general law. For example, a bull-ant can be found in relation to its habitat. In this relational

Table 6.2 Modeling with mediums of painting and collage

Observations	Transcript	Field notes
27.2 Bug machine		*The day before he had found a large bull ant near the sand pit and called for his teacher to come and get it. She had carefully removed it (using a glass and cardboard) to the adjacent bush land whilst he watched and told her about how bull ants have jaws and teeth to bite.*
Christian is at a table with food dye and brushes when he spontaneously paints and explains about a machine he has represented on paper that can suck up bull-ants. The machine he painted represented a functional solution to managing stray bull ants that might bite and offered thought as to what might happen should they get sick	*Christian: It goes up there and it gets the ants and this is when they go to the dentist.* Teacher J: *Go to the dentist?* *Christian: Yeah that's when they get sick and then they go here*	
21.2 Pacman person chomping	Teacher P: *Oh wow…. what fun….* (she plays with the pacman person opening its mouth.)	Later in the day when *Christian's* peer Colleen stamps on a beetle, he cries out loud in anguish. Christian has strongly expressed concerns about preservation of life. Teacher J empathises and begins a new search with a group of children to find a new insect in the yard.
Christian continues to re-present his earlier idea about digestion and has chosen the collage table to create an imaginary bug like pac-man from a round piece of paper. He wants the character to function with a mouth that opens so it can 'burp, eat, bite and chomp'. With encouragement from assistant Teacher P, he cuts a design that allows the character to do this. Teacher P role plays with Christian's creation and he jumps with excitement when it is animated in front of his peer Colleen. Christian often converses with the creatures he finds and is delighted when Teacher P brings this imaginary creature 'to life' with comic voices.	*Colleen: Excuse me….* Teacher P: *He got a circle right and he got two dots for eyes and he cut, cut, cut for the mouth…. Look.*	

model the child sees a specific individual form of a bull-and and an ant hole, in its universal form it is an organism and habitat. The relations between habitat and insect represent knowledge generation that was created historically, as a form of classifying and organizing the world, as a scientific knowledge tradition. Consequently, a concept must reflect the process of its historical and scientific development. That is, the child must also have the opportunity to investigate its environment and notice the bull-ant and ant hole are always linked (specific) with the experience of generating a scientific model of organism and habitat (general). This historical and scientific development can be reflected through the knowledge base of the teacher, as the teacher directs the children's attention to specific features of the ecosystem, or through the strategic use of books and charts which in themselves contain the history of knowledge of Western science.

Although children are not working independently to re-discover bodies of established knowledge, they do engage in an investigative process guided by their teacher, which allows them to build theoretical knowledge and use dialectical thinking for establishing the relational knowledge (in the form of a model) that underpins the historical and scientific journey undertaken initially in revealing the discovery and building of the particular knowledge system – in this case science.

The case example discussed above highlights how the teacher helps Christian to think scientifically about his world. Rather than simply looking for insects in his environment, Christian is supported to think paradigmatically about his everyday world using theoretical knowledge of the ecosystem. Christian has not been left to discover the world on his own. But the valued forms of knowledge that have been constructed by human society to explain why the ant was in the wrong place, were introduced to Christian through the thoughtful interactions of his teacher. This example contrasts with the example of studying potions, where the teacher's belief system about her role and how children learn (individualistic paradigm) created very different conditions for learning for Molly and Lara. The children brought their own personal knowledge of 'cooking' or 'nursery rhymes' and tried to make meaning of the materials through these lenses. Narrative knowledge was supported. But the narratives they formed did not help them to scientifically understand the materials they were playing with. What is important here is recognizing how beliefs about knowledge construction (narrative, empirical and theoretical) and children's development (individualistic, social interactionist or cultural-historical) determine what action a teacher takes. As we see in the examples given above, the consequential outcomes for early childhood science education are very different.

6.6 Knowledge Construction in Early Childhood Science Education

In this chapter we have examined three types of knowledge construction, narrative, empirical and theoretical. Our discussion focused on preschool aged children, and through this it was noted that some early years teachers have difficulties with considering science education from an empirical knowledge construction perspective because most preschool practices within European and European heritage communities, tend to privilege narrative knowledge construction. It was argued that theoretical knowledge construction and dialectical thinking allowed children to engage in historically developed empirical knowledge and turn this empirical knowledge into personal knowledge when:

- children had the opportunity to study the particular bull ant whilst considering at the same time the general concept of an insect as an established body of scientific knowledge;
- insects were studied in relation to the concept of habitat, and food sources, allowing for the development of a relational understanding or concept of a rudimentary ecosystem
- children created a model that represented the essence of what they were studying, so that core elements of the concept could be consciously considered.

Whilst theoretical knowledge allows children to turn scientific concepts into personal concepts, this should not discount the usefulness of narrative knowledge and empirical knowledge construction. These other forms of knowledge construction are drawn upon extensively in preschools, but on their own they have been shown to be less effective forms of knowledge construction for early years learning in the sciences when the teacher is not clear about the core concepts s/he is investigating with children. It is important to be aware of what knowledge construction dominates in a particular institution, such as a family, preschool or school. Knowing what dominates, or that which is found most comfortable for teachers and particular groups of children, means that a more explicit approach to introducing other forms of knowledge construction are necessary. We now turn to the final chapter in this section, where we examine how children are positioned in research and how this influences knowledge construction in science for early years education.

References

Aikenhead, G., & Michell, H. (2011). *Bridging cultures: Indigenous and scientific ways of knowing nature*. Toronto, Canada: Pearson Canada.

Bruner, J. (1986). *Actual minds, possible worlds*. Cambridge, MA: Harvard University Press.

Carter, L. (2007). Sociocultural influences on science education: Innovation for contemporary times. *Science Education*. 1–17. doi: 10.1002/sce.

Davydov, V. V. (1990/1972). Types of generalization in instruction: Logical and psychological problems in the structuring of school curricula. In *Soviet studies in mathematics education* (J. Teller, Trans., Vol. 2). Reston, VA: National Council of Teachers of Mathematics.

Davydov, V. V. (2008). *Problems of developmental instruction: A theoretical and experimental psychological study. International perspectives in non-classical psychology*. New York: Nova Science Publishers.

Fleer, M. (2009a). Supporting conceptual consciousness or learning in a roundabout way. *International Journal of Science Education, 31*(8), 1069–1090.

Fleer, M. (2009b). Understanding the dialectical relations between everyday concepts and scientific concepts within play-based programs. *Research in Science Education, 39*, 281–306.
Fleer, M. (2010). *Early learning and development: Cultural-historical concepts in play*. New York: Cambridge University Press.
Fleer, M. (2011). "Conceptual play": Foregrounding imagination and cognition during concept formation in early years education. *Contemporary Issues in Early Childhood, 12*(3), 224–240.
Hedegaard, M., & Chaiklin, S. (2005). *Radical-local teaching and learning a cultural-historical approach*. Aarhus, Denmark: Aarhus University Press.
Sere, M. G. (1985). The gaseous state. In R. Driver, E. Guesne, & A. Tiberghien (Eds.), *Children's ideas in science* (pp. 105–123). Milton Keynes, UK: Open University Press.

Chapter 7
Positioning Children in Research and the Implications for Our Images of Their Competences

Niklas Pramling

Abstract This chapter illustrates and discusses how to produce valid knowledge about children's concepts in science (understanding, learning, development). The discussion presented revolves around a theoretical presentation of communication and re-analyses of empirical data from published research on children's reasoning and understanding. Through this analysis, the chapter examines how images of children's competences in research are produced. The chapter also examines children's understandings of the interview situation, showing how the 'same situation' is conceptualized differently by the interviewer and the child. In addition, the communicative framing and analysis process are critiqued in order to demonstrate how the research interview, as a common method of research into children's science understanding, needs to be reconceptualised as a social practice, where the collaborative unfolding and meaningful exchange between interviewer and child is foregrounded.

Keywords Interview process • Interview situation • Interview as collaborative social process

7.1 Introduction

In this chapter we will raise, illustrate and discuss an important matter in research – how to produce valid knowledge on children's concepts in science (understanding, learning, development). The discussion will revolve around a theoretical discussion on communication and re-analyses of empirical data from published research on children's reasoning and understanding. In Chap. 1 it was noted that the academic legacy of early childhood science education is grounded in constructivism. Consequently, in this chapter we will write extensively on Margaret Donaldson's famous and important book *Children's Minds* from 1978 in order to bring to the early childhood science education literature, historically important features critiqued in developmental psychology, that we believe shaped the need for new theories for informing contemporary directions in early childhood science education.

M. Fleer, N. Pramling, *A Cultural-Historical Study of Children Learning Science*, Cultural Studies of Science Education 11, DOI 10.1007/978-94-017-9370-4_7

We will also present to an international readership, Karsten Hundeide's book *Piaget i kritisk lys* [Piaget in a critical light] from 1977 that to large extent foreshadowed Donaldson's book, but because it was published in Norwegian has not been as well known internationally as the latter. Introduced are examples from more contemporary research and theorizing on how to conceptualise children's conceptualisation and reasoning about natural phenomena on the basis of interviews.

7.2 Studying and Producing Images of Children's Competences in Research

In many different developmental settings, what Aronsson and Hundeide (2002) refer to as 'examination questions', are commonly found. On the basis of children's responses to such questions, claims are made about their competence, level of development and/or understanding. Hence, how to interpret such data is of pivotal importance to a developmental science as well as to educational practices. In their study, Aronsson and Hundeide re-analysed children's responses to such questions in order to "delineate a type of logic that might explain children's 'immature' responses in terms of their sociability rather than in terms of their default qualities" (p. 174). Taking a dialogical perspective – based on the theoretical writings of Bakhtin and Vygotsky – Aronsson and Hundeide emphasise that testing needs to be understood as a "highly collaborative affair, involving mutual adjustments between experimenter and 'subject'" (p. 175), or interviewer and child. A dialogical perspective thus means to consider sense-making as a collaboratively evolving activity rather than as an expression of the child's thoughts as such, the latter referred to as a monological model of explanation. In developmental science, the critique against a monological perspective on children's reasoning was launched in the 1970s, particularly against Piagetian studies, by Karsten Hundeide (1977) and Margaret Donaldson (1978). In her famous book, *Children's Minds*, Donaldson reports studies where classic Piagetian tasks are reframed. As is well known, Donaldson looks particularly at problems concerning 'decentration', that is, the ability to see something from another's point of view. The inability to do so, commonly understood as 'egocentrism' was found to be integral to a Piagetian model of development. The well known test used by Piaget to study young children's (in)ability to decentre is the 'three-Mountains test' and this was undertaken with children aged under 7 years old and younger. We briefly summarise this work here as a reminder to the reader of what was foundational to science education in the early years, but also to Donaldson's critique. As Donaldson explains, the child is shown a three-dimensional model of mountains and asked to indicate on pictures how the mountains appear from various points of view where a doll is said to look at the mountains. Alternatively, the child is given cardboard mountains and asked to arrange them according to photographs (Donaldson, 1978). What Piaget found when presenting young children with these tasks was that "Children up to the age of around eight, or even nine, cannot as a rule do this successfully; and there is a powerful tendency among children below the age of six or seven to choose the picture – or build the model – which represents their own point of view – exactly what they themselves see" (ibid., p. 19). What Donaldson and her colleagues did was to design situations where children's ability to decentre could be tested in a form that presumably made more sense to the children than the rather abstract situation used by Piaget. Presenting the child with a model of walls and a doll and asked to hide the doll from the policeman's view, and, in a more complex test, hide the doll from the views of two policemen, "The results were dramatic" (p. 22). Giving this task to 30 young children (3.6–5-years old), Donaldson writes, "90 per cent of their responses were correct. And even the ten youngest children, whose average age was only three years nine months, achieved a success rate of 88 per cent" (p. 22). Donaldson argues that this markedly different result from what is reported by Piaget, among other things, can be explained in terms of the fact that the alternative test "makes human sense" (p. 25). Children are generally familiar with playing hide-and-seek. Hence, the problem is presented in a form that children can relate to and have experiences of.

The importance of the child being able to relate to the problem presented to him or her was also pointed out and forcefully illustrated by Karsten Hundeide in his preceding book on critiquing and reframing the Piagetian model of development. In his book, Hundeide describe a number of empirical studies with children where either (i) the question is held constant (i.e., the same question is asked as used by Piaget) and the objects (or cards of some kind) are varied, or (ii) where the question is varied and the objects are held constant (i.e., the same objects as Piaget used, are used). Both forms of variation lead to results questioning Piaget's claims about the logical structures of children's thinking at the various stages of his developmental model. To give only one brief example of a study where the objects were constant and the mode of asking the question differed. Showing children in second grade a number of circles [i.e., what schooled persons would typically refer to as circles (see Luria, 1976); cf. our reasoning in Chapter 12 about the constitute nature of language], six in all, half of which were black and half of which were white (i.e., only a black outline, in succession from left to right from smallest to largest, resulting in every second one being white and every second one being black), two alternative questions were asked: "Kan du sette et kryss på den av de hvite rundingene som er nest minst?" [in English: "Could you put a cross on the second smallest white round one?], while the other group was asked: "Kan du sette et kryss på den av snøballene som er nest minst?" [in English: "Could you put a cross on the second smallest snowball?] (p. 49, italics omitted). It turned out that the children found the second question far easier than the first one. Snowballs where thus interpreted as being more meaningful as a category for these children than "hvite rundingene" [white round things]. This also testifies to the cultural nature of children's experiences. For these Norwegian children, 'snowball' is a meaningful category. But it would hardly be one for children growing up in some other parts of the world, which further strengthen Hundeide's reasoning, and thus, critique of Piaget's model of explanation as culturally biased.

In fact, as summarized by Hundeide (1977), subsequent cross-cultural research on Piagetian-type tasks has clarified that (i) children who have had much contact with Western culture succeed to a higher degree than children who do not have this background, (ii), those who have attended Western schooling succeed to a higher degree than those who do not share this background, and (iii) those children who have grown up in a technology-intensive environment succeed to a higher degree than those raised in rural areas (ibid.). Consequently, Hundeide concludes, a child's intellectual development cannot be understood separate from the socio-cultural experiences that he or she has been allowed to make, the practices participated in (cf. Luria, 1976; Vygotsky, 1987). With this backdrop of critique in mind, we now turn to the interview situation in research for examining how knowledge about early childhood science education is constructed in a more contemporary context.

7.3 The Child's Understanding of the Interview Situation

In Sommer, Pramling Samuelsson, and Hundeide's (2010) book, *Child Perspectives and Children's Perspectives*, Karsten Hundeide's important work on how children perceive interview situations are described and theorized. His so-called 'reconstruction method' is presented. This method allows the child to retell, demonstrate and dramatise his or her experience of the interview. The following procedure is used. The child takes part in an interview, as common practice, carried out by a researcher in an experimental room. After the interview, the child's preschool teacher comes to the room to take him or her back to the other children. While walking back, she asks the child if he or she was given the reward for being so clever during the interview. Since the child has not received the reward, the preschool teacher and child return to the interview room to get it. In the room, all the objects used during the interview, such as bricks, still lay on the table. The child gets a small reward, for example, a piece of chocolate. The reconstruction procedure is now initiated:

> 1. While the child is eating the chocolate, the preschool teacher says, I have never participated in such an event before; maybe you could tell me what happened while you were here together with the man or woman? The child then gives a verbal description of what he or she experienced.
>
> 2. Then the preschool teacher goes further and comments: I see these bricks (from the Piagetian conservation experiments) are still lying there (pointing to the table), could you show me exactly what happened?
>
> 3. When that is finished, the preschool teacher says, Maybe we could try to play together what happened, you can be the man [or woman, i.e., the interviewer] and I will be the child – ok? (Sommer et al., 2010, p. 125)

In this way, the child could tell about, demonstrate and role-play his or her experience of what had taken place during the interview. Among other things, through this procedure, Hundeide was able to show that the children "indicated that they had been participating in a 'guessing game' and that 'they had answered all the questions correctly'", while "Other children produced fantasy stories linked to the bricks in the number conservation experiment – stories about the families that lived there, about the mother who went out shopping to the other 'block', etc." (ibid., p. 127). Faced with the Piagetian-type task of number conservation, one child (6-year-old

Anne), in the role of the interviewer (see above), spread out the bricks into a long row and asked the preschool teacher (in the role of the interviewee, see above), "Is this a snake?"

Preschool teacher:	"Whether it is a snake?"
Anne:	"Yes, it is a snake!"
Preschool teacher:	"Did he really ask you about that?"
Anne:	"Yes."
Preschool teacher:	"Why do you think he asked you this question?"
Anne:	"I don't know." (ibid., p. 126)

Apparently, child and interviewer had not established intersubjectivity (Rommetveit, 1974) in the interview situation, as to what is being asked of the child, in both senses of the term, that is, what the question posed is, on the one hand, and what activity is expected of the child, on the other. While the interviewer may perceive the interview, or take it for granted, as a test of formal, 'school-based' forms of knowing, the child may perceive the activity as a make-believe playful situation.

Not only does the analysis reveal why children's interview responses cannot be taken simply as windows on their intellectual capacities, it in fact also showed examples of so-called 'false positives', that is, that a child may arrive at the 'correct answer' from the wrong premises. The following example, in relation to the number conservation test, can illustrate this point:

Per:	"He asked me whether there lived the same number of people in this house as in the other house" (pointing)
Preschool teacher:	"What did you answer then?"
Per:	"I answered that there lived just as many persons in both blocks because they were of the same size." (Sommer et al., 2010, p. 128)

Hence, rather than focusing on the fact that there is an equal amount of blocks (houses) in the two rows, even though one of the rows is more spread out, the child 'solved the task' by understanding it in terms of people living in the houses (blocks). What Hundeide refers to as the 'meta-communicative framing' of the participants are critical to what they will perceive the point of the activity being, their role in it, and consequently what they consider to be relevant contributions (questions and answers). The 'same situation' is on many occasions very differently conceptualized by the interviewer and the child. What is studied in research interviews is not some 'free-flowing rationality'. Rather what this research illustrates is how reasoning is situated in practices, that is, that we always understand something from a certain perspective, relative to how we perceive the situation and our position (role) in it.

It should be emphasized that our intention with this reasoning is not to argue the case that interviewing should not be a legitimate method of studying children's development. We know a great deal about children's development from interviews. Rather, what we, and others (e.g., Aronsson & Hundeide, 2002; Säljö, 2000; Wallerstedt, Pramling, & Pramling Samuelsson, 2011), argue is that how we interpret such data need to be grounded in contemporary theorizing on the situated nature of human knowing and sense making – or alternatively phrased, the positioning of children in research – as illustrated with reference to classic and more recent studies in this chapter where children's thinking in science is featured.

7.4 Communicative Framing and Analysis

Rommetveit (1985) has made the point that the Piagetian paradigm, or in other terms a monological model, in disregarding the social situation of testing can be seen as what he refers to as a 'negative rationality', that is, explaining children's thinking "in terms of default qualities, what is lacking in their thinking" (Aronsson & Hundeide, 2002, p. 175). This is particularly important to note when considering how knowledge in science has been constructed. This is a point we take up further elsewhere in this second section of the book where we review the literature across a range of cultural contexts. In contrast to constructivist framings in research, the aim of the reanalysis and reconceptualization made by Aronsson and Hundeide is to "move beyond negative rationality toward a description of children's relational rationality" (loc. cit.). In brief, from the latter perspective, "interview responses must be understood in terms *of participation patterns*" (p. 181, italics in original). This means, among other things, that participants in a dialogue tend to align with one another, rather than question the premises of a question or challenge the 'face' of the interviewer, even if the question is an absurd one (e.g., "Is milk bigger than water?", Donaldson, 1978, p. 72). An analytical consequence of this dialogic perspective on meaning is that the research interview, as a common method of research into children's science understanding, needs to be analysed as a social practice (Säljö, 2000). This means that what we have access to and can analyse in an interview is the collaborative unfolding of sense between interviewer and child. Rather than understanding what the child says as his or her thoughts, understanding or conception, the child's utterances need to be analysed as responses to the questions posed, not only as a factual statement but also as a social response. This moves the analytical frame from a constructivist to a cultural-historical perspective. As shown by Aronsson and Hundeide and others, children may at times go to great length to align with an interviewer in order to maintain the conversation, the activity. Contrasting what they refer to as children orienting towards a 'relational rationality' to a 'scientific rationality', as premised in research (based on a monological conception, see above), they argue that the former is guided by the principles of 'participation', 'mutual understanding' and 'alignment' while the latter is guided by an expectation of 'fact finding' and 'logical explanation'. These are markedly different rationalities for guiding actions and participating in an activity, in this case the research interview. As Aronsson and Hundeide argue, "Young children seem to care more about keeping social relations going than about logical consistency. There is thus a greater tolerance for contradictions in relational rationality, in that alignment concerns are more central than fact finding or logical explanations" (p. 183). Returning to the overarching theoretical framework of their analysis and discussion, Aronsson and Hundeide point out that "On a theoretical note, we have corroborated and restated Vygotsky's point that the social level is a primary level in human action" (p. 184).

In a similar vein, Pramling (2006b) argued for the need to analyse the research interview as a social practice. Reanalysing empirical excerpts from one of Piaget's most famous and influential books, *The Child's Conception of the World* (Piaget, 1926/1951), how children speak figuratively and use meta-communicative markers were investigated. Building upon the important studies of Hundeide (1977, 1985), Donaldson (1978) and Aronsson and Hundeide (2002), Pramling (2006b) studied "the manners in which children provides perspectives on what they are saying", arguing that if children indicate that they use language non-literally (i.e., figuratively, metaphorically), "this would have serious implications for this line of research and the conclusions drawn about the children's understanding" (p. 454). There are many ways that people indicate in communication the tensious relationship between what they say and what they mean. In this case, what was analysed was certain verbal markers (Goatly, 1997), such as 'a kind of', 'similar to', 'as if', and 'like'. Simply put, such markers clarify that the speaker does not make a reality claim but that this is rather a manner of speaking, and should therefore not be taken (interpreted) literally. Consider the following as an example; a child is asked the very difficult question, "Do you know what it means to think of something?" (Piaget, 1926/1951, p. 37):

> TANN (8) thinks with his "*mind*". "What is the mind?–*It is someone who isn't like we are, who hasn't skin and hasn't bones, and who is like air which we can't see*". (p. 53)

Facing the communicative and cognitive challenge of clarifying what it means to think, the child speaks about it in terms of something more tangible; as if it were an agent ("it is someone" but "who isn't like we are"). The child also qualifies his reasoning, suggesting that it is "like" something (or someone) while at the same time it "isn't like" something else. The notion of thought is thus communicated about in terms of tentative similarities and differences to something other which is easier to talk about. Through his meta-communicative makers, the child clarifies that what he says should not be taken literally. It is a rather impressive undertaking of the child to qualify his speech in this manner. Disregarding the child's meta-markers, Piaget interprets the child's utterances as indicating the child identifying thought with air (Piaget, 1926/1951).

What the children are interviewed about in Piaget's study (1926/1951) is scientific phenomena, for instance where rain comes from. To give one example, from the chapter on 'the origins of child animism' of a 6-year-old child named Had (text in italics are the child's, text in quotes are the interviewer's and plain text is Piaget's own comments):

> HAD (6) "Can the sun do whatever it likes?–*Yes, because it's alone with the moon*–And the clouds?–*Yes, because they are alone with the other clouds*", etc. The meaning of these words is sufficiently clear from the following answer: "Can you do whatever you like?–*Yes, because my mother sometimes lets me*". (Piaget, 1926/1951, p. 227)

According to Piaget's own analysis of this excerpt, it illustrates how the child "endows all objects with freedom of movement for the reason that they are 'alone', that is to say that no one commands them nor supervises what they do" (loc. cit.). Hence, the child's answers are read as stand-alone claims about reality, how the child thinks something is. If instead interpreting this excerpt in a more dialogical manner (cf. Aronsson & Hundeide, 2002), the first thing to consider is what the child's answers are answers to. The initial question posed by the interviewer, "Can the sun do whatever it likes?" actually constitutes the sun as an agent capable of doing something, the question being whether it can *do* whatever *it likes*. Hence, already the initial question animates the phenomenon spoken about (the sun). As several theoreticians of language and communication have emphasized (e.g., Goffman, 1981; Rommetveit, 1974, 1985; Vološinov, 1929/1986; Wittgenstein, 1953), language does not simply represent reality, rather it is a device for structuring reality and constituting it in interesting and relevant ways for various communicative purposes. Of particular importance to our present discussion is the fact that the interviewer through his initial formulation actually communicatively frames (Goffman, 1974) the phenomenon in animistic terms. Already young children tend to be sensitive to such communicative features and align their speech accordingly (Aronsson & Hundeide, 2002). As seen in the excerpt, the child reasons in terms that are reasonable within the framework established by the interviewer's question, using 'because' as a kind of meta-signal that motivates how the phenomena in question could be spoken about in the suggested way (Pramling, 2006b). Simply put, our argument is that while the interviewer communicatively frames the issue to be talked about in animistic, as-if terms and the child sensitively aligns with and responds in a corresponding fashion, these two features of the interaction are not considered in Piaget's own analysis of the data. Instead, the child's expression is read as indicating a conception held, in this case, that the child thinks animistically. In alternative terms, we argue that the interviewer frames the issue in an as-if manner (as a way of speaking rather than making a claim about the nature of the phenomenon) and while the child responds accordingly, the child's answer is read as indicating an 'undeveloped' way of thinking, while the interviewer's analogous turn of phrase is not considered (or seen merely as a manner of speaking). In a way, the child is thus communicatively framed (in both the Goffmanian sense and in the normative sense of being tricked) by the interviewer into a position where it is unlikely that he or she will give what is considered a 'correct' response to the question (cf. Hundeide, 1977, for a similar and elaborate discussion). With the alternative perspective we have tried to illustrate with reference to Hundeide (1977), Donaldson (1978), Aronsson and Hundeide (2002) and Pramling (2006b), children appear communicatively sensitive and competent rather than cognitively 'insufficient' or 'undeveloped'. As theoretician of science, Norwood Hanson (1958/1981) emphasized in his classic study, *Patterns of Discovery*, there is no theory-neutral way of interpreting data; scientific observation is 'theory-laden'. Whether we take a Piagetian or a Vygotskian perspective when interpreting empirical data, such as interviews, we see different things and importantly we produce very different images of children's capabilities in research (Pramling, 2006a; Wallerstedt et al.,

2011). As researchers, it is important not to make ourselves blind to our own contributions to the knowledge we generate.

In this book we privilege a cultural-historical reading of early childhood science learning and teaching. As such, we believe that a cultural-historical framing of the interview context and the interview itself, is more productive than the traditional approaches used, such as, constructivist or children's science as presented in Chap. 1 and critiqued in this chapter. In the following three chapters, we investigate empirically how some important features of early childhood science education plays out in a concrete sense, focusing on how teachers and children manage issues of representation.

References

Aronsson, K., & Hundeide, K. (2002). Relational rationality and children's interview responses. *Human Development, 45*, 174–186.

Donaldson, M. (1978). *Children's minds*. Glasgow, Scotland: Fontana/Collins.

Goatly, A. (1997). *The language of metaphors*. London: Routledge.

Goffman, E. (1974). *Frame analysis: An essay on the organization of experience*. New York: Harper & Row.

Goffman, E. (1981). *Forms of talk*. Philadelphia, PA: University of Pennsylvania Press.

Hanson, N. R. (1981). *Patterns of discovery: An inquiry into the conceptual foundations of science*. Cambridge, UK: Cambridge University Press. (Original work published 1958)

Hundeide, K. (1977). *Piaget i kritisk lys* [Piaget in a critical light]. Oslo, Norway: Cappelen.

Hundeide, K. (1985). The tacit background of children's judgments. In J. V. Wertsch (Ed.), *Culture, communication, and cognition: Vygotskian perspectives* (pp. 306–322). New York: Cambridge University Press.

Luria, A. R. (1976). *Cognitive development: Its cultural and social foundations* (M. Lopez-Morillas & L. Solotaroff, Trans.). Cambridge, MA: Harvard University Press.

Piaget, J. (1951). *The child's conception of the world* (J. Tomlinson & A. Tomlinson, Trans.). Savage, MD: Littlefield Adams. (Original work published 1926)

Pramling, N. (2006a). *Minding metaphors: Using figurative language in learning to represent* (Göteborg studies in educational sciences, 238). Göteborg, Sweden: Acta Universitatis Gothoburgensis.

Pramling, N. (2006b). 'The clouds are alive because they fly in the air as if they were birds': A re-analysis of what children say and mean in clinical interviews in the work of Jean Piaget. *European Journal of Psychology of Education, 21*(4), 453–466.

Rommetveit, R. (1974). *On message structure: A framework for the study of language and communication*. London: Wiley.

Rommetveit, R. (1985). Language acquisition as increasing linguistic structuring of experience and symbolic behavior control. In J. V. Wertsch (Ed.), *Culture, communication, and cognition: Vygotskian perspectives* (pp. 183–204). New York: Cambridge University Press.

Säljö, R. (2000). *Lärande i praktiken: Ett sociokulturellt perspektiv* [Learning in practice: A sociocultural perspective]. Stockholm: Prisma.

Sommer, D., Pramling Samuelsson, I., & Hundeide, K. (2010). *Child perspectives and children's perspectives in theory and practice* (International perspectives on early childhood education and development, 2). New York: Springer.

Vološinov, V. N. (1929/1986). *Marxism and the philosophy of language* (L. Matejka & I. R. Titunik, Trans.). Cambridge, MA: Harvard University Press. (Original work published 1929)

Vygotsky, L. S. (1987). *The collected works of L. S. Vygotsky, Volume 1: Problems of general psychology, including the volume* Thinking and Speech (R. W. Rieber & A. S. Carton, Eds., N. Minick, Trans.). New York: Plenum.

Wallerstedt, C., Pramling, N., & Pramling Samuelsson, I. (2011). In tune with the learner's perspective in music: Theoretical and analytical considerations when interviewing children. In D. Harcourt, B. Perry, & T. Waller (Eds.), *Researching young children's perspectives: Debating the ethics and dilemmas of educational research with children* (pp. 163–177). London: Routledge.

Wittgenstein, L. (1953). *Philosophische untersuchungen/Philosophical investigations* (G. E. M. Anscombe, Trans.). Oxford: Basil Blackwell.

Part III
The Pedagogical Construction of Learning Science

Chapter 8
Learning and Metaphor: Bridging the Gap Between the Familiar and the Unfamiliar

Niklas Pramling

Abstract In this chapter metaphors and related figures of speech are shown to be necessary and integral parts of sense making and learning. It is shown how this kind of speech has been studied in research in relation to children's understanding and learning. It is argued that different methodological approaches have led to different notions of children's abilities. Two recent studies on children and metaphorical speech in science-learning activities in preschool are introduced. One study investigated the nature and use of such speech and another study looked at a particular form of metaphorical speech, anthropomorphism. It is argued that metaphor is one way of establishing relations between different things without collapsing them into one and the same.

Keywords Meaning making • Metaphor • Methodology

8.1 Introduction

This chapter will focus more in-depth on a theme that we touch upon throughout chapters of this book – how learners and teachers in speech relate something novel and only partly known to something more familiar. After introducing the idea that metaphors and related figures of speech are necessary to, and integral parts of, sense making and learning, we will discuss how this kind of speech has been studied in research on children's understanding and learning. We argue that different methodological approaches have led to different notions of children's abilities. We then summarise two recent studies on children and metaphorical speech in science-learning activities in preschool, one study investigating the nature and use of such speech and another study looking at a particular form of metaphorical speech, anthropomorphism. Finally, some more overarching conclusions are drawn.

Metaphors and similes are central to verbal actions, relating the familiar to the novel, but there are also other forms of speech such as analogies that are used in this way. The empirical foundation for this chapter is Pramling's work on metaphor in early childhood science education, which is one of a limited amount of studies looking at this communicative feature with younger children. There is more research

M. Fleer, N. Pramling, *A Cultural-Historical Study of Children Learning Science*, Cultural Studies of Science Education 11, DOI 10.1007/978-94-017-9370-4_8

into metaphor in science education with older students (e.g., Aubusson, Harrison, & Ritchie, 2006; Bishop & Anderson, 1990; Dagher & Boujaoude, 2005; Halldén, 1988; Pedersen, 1992; Tamir & Zohar, 1991). Only more recently has this empirical interest been investigated in early childhood education. How young children and their teachers use metaphors when learning about nature, that is, how they use what they already know to learn something new, without solely reducing the novel to the familiar, and thus not learning anything qualitatively new, but merely confirm what he or she already knows, are important questions to educational research. Metaphor is one way of establishing such relations between different things without collapsing them into one and the same.

A common strategy used when people – children as well as adults – try to make sense of, or communicate about, something unfamiliar is to speak about the novel in terms of the more familiar. We can analyze this communicative act though attending to the metaphors used and, more specifically, how they are used to 'bridge the gap' between the unfamiliar and the familiar. As Wertsch (1998) has emphasized, to appropriate a cultural tool often requires an extensive familiarization process; we do not simply take over in any straightforward manner, once and for all, a cultural tool. Rather, we become increasingly familiar with how to use a tool in relevant and flexible ways in various practices. The use of metaphor when starting to make sense of something unfamiliar in more familiar terms can provide insight into this process of appropriation. This process could be studied on a collective level of the formation of scientific knowledge (Keller, 1995, 2002; Ochs, Gonzales, & Jacoby, 1996; Pramling & Säljö, 2011) as well as on an individual level. The metaphorical nature of sense making and communication is particularly apparent when people encounter more abstract forms of knowledge, such as scientific knowledge. To give a few examples; a child facing the challenge of making sense of the ozone layer may speak about it in terms of a sheet (Cameron, 2003), while a geneticist explaining his or her field of expertise to a lay audience may talk about 'code', 'letters', and 'translation' (Knudsen, 2003; Pramling & Säljö, 2007).

8.2 Studying Metaphorical Speech and Changing Notions of Children's Abilities

The interest in metaphor has a long tradition, going back to the writings of Aristotle in Greek Antiquity (Aristotle, version 1999, version 2000). For a long time, metaphor was considered a particular kind of speech for ornamental and/or rhetorical purposes (for historical accounts of metaphor, see e.g., Draaisma, 2000; Leary, 1990; Roediger, 1980). In more recent times, the interest in psychology, education, linguistics and other fields of study, was renewed with the influential book, *Metaphors we Live by*, written by George Lakoff and Mark Johnson. Since the publication of their book in 1980, many studies have shown how metaphors play important parts in human sense making and communication.

Simply put, using a metaphor means to speak about something, typically less familiar, in terms of something else, that it in a literal sense is not (Lakoff & Johnson, 1980). Phrased differently, we can say that metaphor is the process through which we use our primary tool for learning – language – in functional ways for speaking about and making sense of a changing world of experiences. If we look at metaphor in this way, then learning to speak metaphorically and understand such speech become important features of a child learning a language and learning about the world through language. In a recent account of children's development, developmental psychologist Stephen von Tetzchner (2005) argues that:

> The role metaphor has in language makes the understanding and use of metaphors the most important developmental aspect of language in school-age children and adolescents. To understand a word in both a literal and a transferred sense is an ability that has just started to form at the age of 5-6 years and that appears to receive a burst in development during adolescence. (p. 345; our translation)

The importance for the language development of the child here ascribed to the use and understanding of metaphors is clear. Departing from this reasoning, there are important features of how we look at children's abilities that we would like to comment on. The first concerns what is implied as a relevant criterion of a child's abilities in this regard. Previous research into children's 'metaphoric abilities' (Knowles & Moon, 2006) has primarily been laboratory-based investigations when children are faced with the problem of explaining the rationale of metaphorical utterances presented by the experimenter. There has been much critique against such studies. This critique has mainly focused on two points. First, that the situation, where the child is presented with a-contextual utterances in an unfamiliar environment, is problematic (see also, Chap. 7); Second, what is taken as an indicator of the child's understanding may need to be reconsidered (Cameron, 2003; Pramling, 2006). It can be argued that taking the child's explanation of a metaphor as an indicator of her understanding conflates two different forms of knowing; a knowing in use and a meta-knowing. What is asked of the child is to provide meta-knowledge. In fact, it is often difficult even for adults to clarify the rationale of certain metaphors. This difficulty does not prevent people from using metaphors in functional and relevant ways in their everyday communication. On the basis of this reasoning, it is important to study children's metaphorical speech in everyday activities, when they engage with other children and/or teachers about, for example, natural phenomena.

In response to the critique raised against previous studies of children and metaphor, in the project from which the examples of this chapter come, everyday conversations between children and teacher and between children around natural phenomena were analyzed. Metaphor is therefore seen as language in use, rather than as a cognitive problem to be solved in the abstract. In the stated study, teacher-led activities in the domain of nature (science) have been documented with video. Themes about nature have been followed from initiation, over consecutive occasions to completion. In this chapter, we will use some of the transcribed excerpts to illustrate our reasoning. The aim of the overarching project on children and metaphoric speech was to investigate the following issues: What kind of metaphors do children and teachers use during activities on natural phenomena (science) in early childhood education, and How are these metaphors used and do the participants indicate how they themselves understand these and how they intend others to understand these utterances?

8.3 Using Metaphorical Speech in Early Childhood Science Activities

Summarizing the findings of the empirical studies, it was found that:

1. Already in early childhood education, teachers as well as children use metaphoric (and other figurative) speech when speaking about natural phenomena. This has previously been seen in studies with older children learning science in school (Cameron, 2003; Jakobson & Wickman, 2007), but not with younger children.
2. There is a rich repertoire of figurative speech in these activities, including analogy, simile, and verbal and gestural metaphors, including animistic and anthropomorphic ones (see below).
3. Such speech appears as a multi-functional tool, that is, children and teachers do many different things with such utterances, such as describing the appearance of something observed and how it differs from something else, explaining and visualizing abstract phenomena and processes, explain other terms, and to mitigate potentially disturbing findings.
4. Some of these utterances are negotiated between teacher and children, but in other cases the conversation proceeds smoothly without the need for explicit clarification of terms (Pramling, 2010).

To just give a few brief examples (we will give more extensive empirical examples of metaphorical reasoning below and in Chaps. 9 and 10 in this book): When encountering a dead shell, this find is spoken about in terms of it being "flat as a pancake" (in a strict sense a simile, but this is not an important distinction to our present discussion). Another example is when they find a plaster in the soil, and speak about this in terms of "What do you imagine the soil thinks when a plaster turns up"? and "What do you think the worm thinks when he crawls onto an old plaster?" In the first case, something inanimate (soil) is spoken about as if it were animate and an intentional agent and in the second case the worm is made into a cognizant (male) being, concerned with a plaster (which could here be seen as a form of anthropomorphism). Through these utterances, the teacher engages the child in thinking about natural and biodegradation (and non-biodegradable) finds.

The prevalence of a particular form of metaphoric speech was observed, so called anthropomorphic speech, that is, speaking about the non-human world in human terms. The occurrence of such speech in children has long been known. It was pointed out already by Piaget (1923/1926, 1926/1951) as characteristic of children's thinking (and speech) as well as by Susan Carey (1985) in her work. Given the apparent prevalence of anthropomorphic speech in conversations about natural phenomena, Thulin and Pramling (2009) reanalyzed empirical data in the form of transcriptions of recordings of a prolonged theme-work in preschool on ecology (Life in the tree stump).

8.4 Giving Nature Human Form

Against the background of previous accounts of anthropomorphism as characteristic of children's thinking (Carey, 1985; Piaget, 1923/1926, 1925/1951), the following issues were investigated: Is there any pattern in the use of such speech; Is such speech introduced by children and/or teachers; and how is such speech responded to by the interlocutors (children and teachers)?

Summarizing the findings, it was observed that:

1. Anthropomorphic speech was primarily used to speak about animals (their conditions, appearance and behavior).
2. Of a total 128 anthropomorphic utterances, 24 were made by the children and 104 by the teachers.
3. At times, the children respond in line with such speech, as established by the teacher, but on other occasions even these children as young as 4–6 years, questioned the teacher using such speech (Thulin & Pramling, 2009).

The nature of anthropomorphic speech in the activities revolving around a tree-stump and what was found in and adjacent to it, can be illustrated with a few examples. One find was a shell. The following exchange between one of the teachers and a child (4 years, 9 months old) ensues:

Teacher: You mustn't touch it, because you'll frighten it Disa, won't you?
Disa (4.9): It has to come out.
Teacher: Yes, it has to, but then you must be careful. Maybe you can talk to it. (Thulin & Pramling, 2009, p. 143)

Suggesting that the child speak to the shell to make it come out, constitutes the animal as a communicable agent much like a human being, responding with an understanding of human speech. It is important not to ridicule the teacher or see her utterance as incorrect; through her speech she makes the child attend to something of great importance in learning about nature, to handle animals (and in extension, engage with nature) in a responsible and careful manner. An important socialization takes place through such conversations.

Another example of speaking about nature in human terms is the following exchange between a teacher and several children observing a woodlouse found in the tree-stump:

Max (6.8):	It's landed upside-down.
Teacher:	How many legs has a woodlouse actually?
Isa (4.5):	It's got all its side full.
Teacher:	Yes, try and count them Isa, you've got lots of them there.
Isa:	Ten.
Teacher:	Ten legs!
Disa (4.9):	Ye-es.
Teacher:	Imagine if we'd had ten legs, what would it have looked like?
Lars (5.2):	It wouldn't have looked – I've got two.
Teacher:	You've got two legs, yes. Imagine if we'd had ten legs, imagine needing shoes for all ten legs – feet.
Disa:	Hmm.
Teacher:	We need shoes when it gets colder, don't we? Wonder if woodlice need shoes?
Carl (6.2):	No.
Teacher:	What do they do to get warm, then?
Carl:	They put inside to get warm.
Lars:	Don't think so, I think they put their hands inside the shell.
Teacher:	Inside the shell?

[---]

Teacher:	Do you know what we're talking about Disa? We're talking about if it gets cold for these woodlice, what do they do then? We put on our winter shoes, don't we?
Carl:	They go inside the stump.
Lars:	No, they go inside the shell.
Teacher:	Is it warm there then?
Lars:	I think they go inside the shell.

Teacher (turning towards Lars)

Lars:	And warm themselves there.
Teacher:	And warm themselves there, like a quilt, you could say.
Lars:	Like a tortoise does.
Disa:	Snail.
Teacher:	Snail. (ibid., p. 143f.)

During this conversation with the group of children, the teacher makes an initial analogy between the woodlouse and people (the children themselves). Through further prompting the children to consider the need to get shoes for the woodlouse's feet, the teacher directs the children's attention to the question of how these animals keep warm when it is cold. Hence, using human terms, the children are invited to use familiar experience and knowledge to start thinking about this issue. The teacher clearly marks out her own speech as non-literal, using terms such as "imagine", "like" and "you could say". The children come to engage in this thought experiment, suggesting additional examples, "tortoise" and "snail". In this way, human terms and experiences become resources in 'bridging the gap' between something familiar and something less familiar that the teacher engages the children to start thinking about.

The study conducted by Thulin and Pramling (2009), as here briefly summarized and illustrated, gives a radically different image to the previous studies (see above) on children and anthropomorphism. Rather than placing such speech (or thinking) with children, this study primarily locates anthropomorphism with the teachers. It appears reasonable to conclude that children learn to speak in this manner much in the same way as they learn to speak in other terms and genres, by engaging with, and listening in to, others speaking in such a way. Anthropomorphism also appears as a mode of speaking rather than simply an expression of an underlying mode of thinking. Our images of children's capabilities is always 'theory-laden' (Hanson, 1958/1981); there is no neutral way of mapping someone's abilities (see Schoultz, Säljö, & Wyndhamn, 2001, for an elaboration of this discussion; see also this volume, Chap. 7, where we more in-depth discuss this issue). A reason for the contrary findings of Thulin and Pramling (2009) to previous studies is likely that how children were studied differed between these studies (cf. above). It is important to realize that anthropomorphic, and other forms of metaphoric, speech is used in different ways, some which makes possible 'bridging the gap' between what is familiar and what is less familiar; others making conversation stay in the human realm and not giving children access to new ways of conceptualizing nature. Speaking anthropomorphically (metaphorically) is not in itself prolific or limiting; it can be used in ways that develop as well as constrain children's understanding. The study reviewed in this chapter also illustrated the importance, when taking a cultural-historical point of view, of studying cultural tools in use (Wertsch, 1998), rather than as stand-alone objects. In the next chapter we build upon the dialectical relations between familiar and unfamiliar events, by exploring the concepts of simile and metaphor through examining the functional and culturally relevant ways they contribute to scientific thinking.

References

Aristotle. (version 1999). *Poetics* (S. Halliwell, Trans.). Cambridge, MA: Harvard University Press.

Aristotle. (version 2000). *The "art" of rhetoric* (J. H. Freese, Trans.). Cambridge, MA: Harvard University Press.

Aubusson, P. J., Harrison, A. G., & Ritchie, S. M. (Eds.). (2006). *Metaphor and analogy in science education*. Dordrecht, The Netherlands: Springer.

Bishop, B. A., & Anderson, C. W. (1990). Student conceptions of natural selection and its role in evolution. *Journal of Research in Science Teaching, 27*(5), 415–427.

Cameron, L. (2003). *Metaphor in educational discourse*. London: Continuum.

Carey, S. (1985). *Conceptual change in childhood*. Cambridge, MA: MIT Press.

Dagher, Z. R., & Boujaoude, S. (2005). Students' perceptions of the nature of evolutionary theory. *Science Education, 89*, 378–391.

Draaisma, D. (2000). *Metaphors of memory: A history of ideas about the mind* (P. Vincent, Trans.). Cambridge, UK: Cambridge University Press.

Halldén, O. (1988). The evolution of the species: Pupil perspectives and school perspectives. *International Journal of Science Education, 10*(5), 541–552.
Hanson, N. R. (1981). *Patterns of discovery: An inquiry into the conceptual foundations of science*. New York: Cambridge University Press. (Original work published 1958)
Jakobson, B., & Wickman, P.-O. (2007). Transformation through language use: Children's spontaneous metaphors in elementary school science. *Science & Education, 16*, 267–289.
Keller, E. F. (1995). *Refiguring life: Metaphors of twentieth-century biology*. New York: Columbia University Press.
Keller, E. F. (2002). *Making sense of life: Explaining biological development with models, metaphors, and machines*. Cambridge, MA: Harvard University Press.
Knowles, M., & Moon, R. (2006). *Introducing metaphor*. London: Routledge.
Knudsen, S. (2003). Scientific metaphors going public. *Journal of Pragmatics, 35*(8), 1247–1263.
Lakoff, G., & Johnson, M. (1980). *Metaphors we live by*. Chicago, IL: University of Chicago Press.
Leary, D. E. (1990). Psyche's muse: The role of metaphor in the history of psychology. In D. E. Leary (Ed.), *Metaphors in the history of psychology* (pp. 1–78). New York: Cambridge University Press.
Ochs, E., Gonzales, P., & Jacoby, S. (1996). 'When I come down I'm in the domain state': Grammar and graphic representation in the interpretive activity of physicists. In E. Ochs, E. A. Schegloff, & S. A. Thompson (Eds.), *Interaction and grammar* (pp. 328–369). Cambridge, UK: Cambridge University Press.
Pedersen, S. (1992). *Om elevers förståelse av naturvetenskapliga förklaringar och biologiska sammanhang* [Students' understanding of scientific explanations in relation to biological contexts] (Studies in education and psychology, 31). Stockholm: Almqvist & Wiksell.
Piaget, J. (1926). *The language and thought of the child* (M. Warden, Trans.). New York: Harcourt, Brace. (Original work published 1923)
Piaget, J. (1951). *The child's conception of the world* (J. Tomlinson & A. Tomlinson, Trans.). Savage, MD: Littlefield Adams. (Original work published 1926)
Pramling, N. (2006). *Minding metaphors: Using figurative language in learning to represent* (Göteborg studies in educational sciences, 238). Göteborg, Sweden: Acta Universitatis Gothoburgensis.
Pramling, N. (2010). Unearthing metaphors: Figurativeness in teacher-child talk about soil and related matters. *Early Childhood Education Journal, 38*(1), 57–64.
Pramling, N., & Säljö, R. (2007). Scientific knowledge, popularization, and the use of metaphors: Modern genetics in popular science magazines. *Scandinavian Journal of Educational Research, 51*(3), 275–295.
Pramling, N., & Säljö, R. (2011). Metaforik för minne och lärande. I R. Säljö (Red.), *Lärande och minnande som social praktik* (s. 42–63). Stockholm: Norstedts.
Roediger, H. L. (1980). Memory metaphors in cognitive psychology. *Memory & Cognition, 8*(3), 231–246.
Schoultz, J., Säljö, R., & Wyndhamn, J. (2001). Heavenly talk: Discourse, artifacts, and children's understanding of elementary astronomy. *Human Development, 44*, 103–118.
Tamir, P., & Zohar, A. (1991). Anthropomorphism and teleology in reasoning about biological phenomena. *Science Education, 75*(1), 57–67.
Thulin, S., & Pramling, N. (2009). Anthropomorphically speaking: On communication between teachers and children in early childhood biology education. *International Journal of Early Years Education, 17*(2), 137–150.
von Tetzchner, S. (2005). *Utvecklingspsykologi: Barn- och ungdomsåren* [Developmental psychology: Childhood and youth] (I. Lindelöf, Trans.). Lund, Sweden: Studentlitteratur.
Wertsch, J. V. (1998). *Mind as action*. New York: Oxford University Press.

Chapter 9
Simile, Metaphor and Learning to Perceive the World in Functional and Culturally Relevant Ways

Niklas Pramling

Abstract In this chapter, we use original data from early childhood education to illustrate, analyse and discuss an important distinction in learning, what something *looks like* and what it *is*. The difference and relation between these two claims, we argue are of great interest when studying learning and it emphasizes how our perceiving is mediated by cultural tools. Managing this distinction could be conceptualized in terms of the dialectics between everyday and scientific concepts. Several examples from research are used to show how metaphors and simile can be used for teaching science concepts to young children. It is shown how the analysis of the data has emphasized the important distinction and relationship between what something *looks like* and what it *is*. That is, children need to be able to discern the object or phenomenon represented (modeled, illustrated) from the mode of representing it. The communicative frame established by the teachers leaves space for children's playful similes. How children perceive the phenomenon being observed, and how they engage with this description is discussed in this chapter. Two or more people sharing attention on a third area is fundamental to education. To establish in speech how to mediate or represent what is observed is one way of coordinating perspectives with this end goal in mind.

Keywords Deictic references • Making crystals • Representations of objects and concepts • Microscopic images

9.1 Introduction

In this chapter, we use original data from early childhood education to illustrate, analyse and discuss an important distinction in learning, what something *looks like* and what it *is*. The difference and relation between these two claims, we argue, are of great interest when studying learning and it emphasizes how our perceiving is mediated by cultural tools. Managing this distinction could be conceptualized in terms of the dialectics between everyday and scientific concepts.

In their important study of preschool in three cultures, China, Japan and the United States, Tobin, Wu and Davidson (1989), among many other things, represent

M. Fleer, N. Pramling, *A Cultural-Historical Study of Children Learning Science*,
Cultural Studies of Science Education 11, DOI 10.1007/978-94-017-9370-4_9

a piece of conversational data that is particularly interesting to the topic and discussion of the present chapter. From a preschool in the US the following observation is made:

> Cheryl [the preschool teacher] then leads the class in an activity involving a felt board and cut-out flannel shapes. Each of the children is called on one at a time to come forward and select a white piece of flannel background. Cheryl explains, 'This blue board is the sky and the white shapes are clouds. Put a cloud on the sky and tell us what the cloud looks like.'

Lisa (in a whisper):	A bird.
Cheryl:	Speak louder, Lisa, so everyone can hear you.
Lisa:	A bird.
Cheryl:	The cloud looks like a bird? [To the class] What do you think? Do you think it looks like a bird? Yes, it does. Good. Thank you, Lisa.
Mike:	This is a cloud.
Cheryl:	Yes, it's a cloud. What does your cloud look like, Mike?
Mike:	Like a cloud. (Tobin et al., 1989, p. 129)

This practice of intentionally pedagogically promoting children's ability to perceive something in terms of something else, that is, through simile, may, Tobin et al. suggest, be culturally variant. In their ensuing conversation with the preschool teacher leading the aforementioned activity, she gives the following rationale of the activity:

> The idea of this activity is to teach children the concept of simile. I gave the children an example of the pattern: 'This cloud is like a da-da-da.' Then they each had their chance. I was less concerned here with what they thought the cloud looked like than with making sure they had the concept of something being like another thing without being the other thing. It's a trickier concept for some kids than others. (Tobin et al., 1989, p. 148; italics omitted)

In cultural-historical parlance, the referred activity illustrates the principle of semiotic mediation (Wertsch, 2007) and how we learn to perceive the world in terms deemed relevant and interesting from the prevailing culture's or institution's point of view. A child looking up at the sky at night (had she been allowed to be awake looking at the sky at night) would not have seen 'The Sign of the Southern Cross' and 'The Great Bear' etc. In order for the child to learn to perceive the night sky in these ways someone needs to support her in 'pointing out and linguistically informing her experience' (Pramling & Pramling Samuelsson, 2011). What something is perceived as like may come to be institutionalized in a culture into what something 'is'; 'That is the Southern Cross', for example. Such a process of institutionalization in a culture could be studied in a historical perspective. The transformation from the tentative 'is like' to the affirmative 'is' of scientific knowledge formation was clarified in the classic study by Ludwik Fleck (1935/1979). In the present chapter, in contrast, we will study how this distinction comes into play and is managed in teacher-child talk around natural phenomena. That is, in the present book and chapter, the relationship between what something 'is like' and what it in institutional terms from a certain perspective 'is' will be studied in terms of the everyday practice in a preschool when they talk about natural phenomena. In passing, we may note that the child in the above excerpt answering that it looks 'like a cloud' is also using a simile (as does the child answering that it looks like a bird); the object on the board is not literally a cloud. However, within the communicative premise as established by the teacher, that the object represents a cloud should be taken as given, and then the task is to clarify what this cloud looks like. What are and are not relevant terms to perceive something in is, in an educational practice, contingent upon the communicative framing (Goffman, 1974) of the activity. Whether it is open for the children to play with how to perceive something or if some kinds of similes are expected and valued in a certain situation is in itself something for children to identify as they partake in a practice.

9.2 What It Looks Like or What It Is: Different Pedagogical Principles and Their Possible Institutional Embeddedness

Asking a child what something looks like (sounds like in music etc.) or what it is, has been suggested by Shirley Brice Heath (1996/1983; cf. Winner, 1988) to be an important difference between the institutions of preschool and school (in the US). Whether this is also the case in, for example, Sweden and Australia some 30 years later, is in itself a question worth investigating empirically. To our understanding these two questions – what something *looks like* or what *it is* – are very different. From a pedagogical point of view the former appears to be far more productive in promoting the child's learning. Asking a child what something looks like allows him or her to use her previous experience and language as resources in making sense of novel phenomena. This also allows the teacher to confirm and thus acknowledge the child's knowledge and, in addition, introduce the child to a new tool for conceptualizing the observation (for an empirical example of this, see Pramling & Pramling Samuelsson, 2010). In contrast, asking a child what something is, while being more distinct, disconnects the novel from the child's previous experience. This question also works summative rather than formatively, that is, it works in checking whether the child has a certain knowledge or not, but does not support the child in furthering his or her understanding. Consequently, scrutinizing our data sets in terms of this distinction between 'what something looks like' and 'what it is' appears to be a worthwhile analytical endeavor. From a pedagogical perspective, particularly interesting is how the tension between what something is like and what it in a conventional sense from a particular perspective is, is managed in teacher-child interaction around, for example, natural phenomena.

9.3 Telling and Explaining

In this chapter we will analyse empirical data from one lesson in a Swedish preschool class. What is referred to as the preschool class is an intermediate form of schooling for the 6-year-olds, intended to bridge between the traditions of preschool and school and thus between play-based and teaching-based ways of organizing learning (for a presentation of this form of schooling, see Pramling Samuelsson, 2006).

The lesson begins with the group of children and their two teachers sitting in a circle. In the following transcripts, the teachers' (fictious) names are written in UPPER-CASE LETTERS. One of the teachers shows an object (a glass jar with something in) and asks, `Does anybody wants to tell about this?` One of the children, Philip, raises his hand and is given the communicative floor. However, he mumbles and it is not possible to make out from the recording what he initially says:

Philip:	[mumbles]
TINA:	There has been water in this, has there been anything else?
Philip:	Salt and the salt has like…
TINA:	What has the salt done?
Philip:	Climbed up the walls and…started to become like warm.
TINA:	It has become warm. What has happened to the water then?
Philip:	It turns into steam.
TINA:	Yes. It has evaporated, yes.
Philip:	Then it's turned into like ice crystals.
TINA:	Yes.

Philip's explanation is interesting for several reasons. He says that, `salt and the salt has like…` This marker (`like`) signals that he has a vague idea about this, but cannot really clarify what it is yet. Asked by the teacher, `what has the salt done?`, which is a kind of question that implies a narrative elaboration (the salt as an agent doing something, an action; cf. Bruner, 2006; Pramling & Ødegaard, 2011), Philip responds that it has `climbed up the walls and… started to become like warm`. He uses this active metaphor of `climbed` to give an explanation, an explanation that he however meta-communicates through his markers (`to become like warm`) he knows is not in a strict sense how it is. Hence, he signals that he speaks in an *as-if*, rather than in an *as-is* manner (see Pramling, 2006, for an elaboration). In her follow up, the teacher connects to the temperature and the water rather than the salt, first stating, `It has become warm`, and then asking, `mm, what has happened to the water then?` Philip responds that `it turns into steam`, which the teacher confirms and reformulates into, `yes. It has evaporated, yes`. Finally Philip adds, `then it's turned into like ice crystals`, and in this way, similar to his previous statements, implies that he has some notion of this process but cannot yet quite explain it, comparing what he sees to `ice crystals`.

Asked if anyone else `wants to tell something else about this`, Magnus raises his hand:

Magnus:	[Walks up to the jar] Yeah, because the water has all turned into st…steam, the salt has also gone up with the steam and it has become a salt stone. [Goes back]
TINA:	Would anyone else like to say anything about this? Have I forgotten anything?
LISA:	I don't underst…
TINA:	Wait Valdemar.
Valdemar:	The salt turned into a salt crystal.

TINA:	Yes, they have become salt crystals.
Valdemar:	Mm.
TINA:	It's a beautiful name, isn't it. Mm. Salt crystals. Yes, mm. And I noticed that Lisa put up her hand.
LISA:	Yes, do you know what. We're going to have a look at them, and they're really beautiful in the jar and on the string, aren't they?
Children:	Mm mm.
LISA:	Yes they are. But we'll see if we can look in this to make it a little bigger. Then we can see it on a board on the wall.
Children:	Mm!

Magnus's explanation is brief and to the point, yeah, because the water has all turned into st...steam, the salt has also gone up with the steam and it has become a salt stone, after which he returns to his seat. Another child adds that the salt turned into a salt crystal (cf. above, like ice crystals), which the teacher confirms with a somewhat ambiguous statement, yes, they have become salt crystals; they perhaps referring to grains of salt. The teacher also adds that It's a beautiful name, isn't it. Mm. Salt crystals. This introduces something that will recur throughout the lesson, aesthetic value judgments. We will return to this issue. Finally, she mentions the next activity to be undertaken; they are to look at the salt crystals under a magnifying glass projected on a whiteboard. For this they have to walk through school to another classroom.

LISA:	Because Marie [a third teacher] has a microscope there.
Boy:	Oh cool!
LISA:	And she has a big board on the wall called a smartboard.
Child:	Oh, and my....
LISA:	And you can see it.
TINA:	Mm.
LISA:	What they look like magnified.

This prompts one of the children to declare, Oh cool! The children are much enthused throughout the lesson (as will be seen below).

9.4 Under the Microscope, Up the Wall: What It Is They Look at, How It Looks, and the Aesthetics of Perception

Having walked to the new classroom and taken a seat, the light in the room is shut and the image from the microscope is projected onto the whiteboard for all to see.

Child:	What's that, is it dust?
TINA:	It looks a little like it.
JOHN:	We can't see anything there now.
TINA:	Jonas. There's nothing there now, but what John is pottering about with now are the salt crystals that have grown in our jar. The white stuff, mm.

JOHN: Do you know what. I've taken a bit of this now, you know better than me what it is because I've got no idea.
Child: Salt crystal.
JOHN: Salt, oh right.
TINA: Not…Lukas.
JOHN: Shall we see what it looks like. [It starts to move on the screen] If we just magnify this a little.
Child: There's the salt crystal. [The crystal can be seen on the screen]
TINA: Oh, get a beautiful picture now, please.
JOHN: Yes, I'll just try to get it to stay still.
TINA: It should glitter.
TINA: Wow! It is glittering! What…what are they doing?

Simply watching the emerging pattern on the whiteboard prompts a child to pose a question; `what's that, is it dust?` In all representational practices (modeling, illustrating, exemplifying etc.), the issue of discerning what is a feature of the phenomenon observed (referred to) from what is a feature of the representational media as such is pivotal. In this case, the child ponders over whether what is observed is the phenomenon (salt crystal) or an 'artefact' in the sense of `dust`. What is in fact observed at this initial time is apparently not clear to the teacher either, as seen in her response; `it looks a little like it`. The teacher clarifies that what they will see is `the salt crystals that have grown in our jar. The white stuff.` A fourth teacher, John, who is also present and assists with the microscope takes the uninitiated role, `I've taken a bit of this now, you know better than me what it is because I've got no idea`, triggering the children to take the role of more knowledgeable and giving contributions such as `salt crystal` in clarifying what John's deictic expressions `this` and `it` refer to. Responding to deictic referencing in terms of verbalization of conventional names is otherwise often done by teachers in learning situations (as we have many examples of in other chapters of this book). Magnifying what is observed, a child immediately points out `there's the salt crystal` as it appears on the whiteboard. Apparently this looks discernibly different to what was previously seen on the screen, which was perhaps dust. The teachers in contrast respond by aesthetic judgments (Jakobson & Wickman, 2007), `Oh, get a beautiful picture now, please`, `it should glitter` and `Wow!`

John says that he can take photos of what they see and print these. Looking at the image on the whiteboard the children start describing what they see:

Child: Oh! It looks like a rock, with a little bridge going over it. [laughs] it looks like stairs!
JOHN: I'll zoom in even more and we'll see what the rest look like…Now it's gone dark…this is now twenty times magnification.

[The children talk at the same time, indistinguishable]

TINA: Shh!
Child: A rock...
TINA: Listen to John now.
JOHN: This is now forty times magnification.
Child: Wow. What is it?
TINA: What can it be? It looks a little like moisture.
LISA: Yes, just like, yes...water.
TINA: Yes, it does, doesn't it.
JOHN: That is a tape...a piece of tape, something catching on a piece of tape, I think.
Child: Oh!
TINA: What do you mean, a piece of tape?
LISA: You had a piece of tape and put the salt on it.
JOHN: Yes, to get it to stay still.
TINA: Oh yes.

Watching the visual patterns on the screen, the children describe what it looks like, `Oh! It looks like a rock, with a little bridge going over it. [laughs] it looks like stairs!` These similes and the child's engagement seem to be triggered by the aesthetic response (`Oh!`). John continues to zoom in on the salt crystals. One child proposes `A rock...`, while another when 40 times magnitude comes into focus exclaims, `Wow. What is it?` Responding, `what can it be? It looks a little like moisture`, one of the teachers, rather than simply stating that they look at the salt crystals, repeats the question giving the children the opportunity to give suggestions. More productive than saying *what it is*, however, at this instance seem to be to encourage the children to describe what they see on the whiteboard *as*, the teacher herself suggesting, `it looks a little like moisture`. The other teacher connects with `yes, just like, yes...water`. However, as suggested by John's comment, it may not be entirely clear what they are in fact looking at, `that is a tape...a piece of tape, something catching on a piece of tape, I think`. This suggestion poses some surprise to the teachers but they soon coordinate their talk. Hence, again the issue of the critical importance of distinguishing between the representational media and the phenomenon comes up for negotiation between the children and teachers (cf. above).

TINA: Shall we see if it is possible without tape.
LISA: Yes...because it might be a bit confusing.
TINA: Yes. It is. Really.
JOHN: Shall we see what this looks like.
TINA: Ah, but it looks like...
LISA: Oh, look!

[it comes into focus and the crystal can clearly be seen]

Children: Ooh!

[Somebody claps their hands]

TINA: Here it comes!

```
LISA:    Oh.
JOHN:    But now it is very...uh...one part is very high up and
         one part is a long way down, so we have to focus
         on it at the same time.
TINA:    But it looks like a crystal.
JOHN:    Shall we see if we can find a good place.
Child:   Yes. Did you see, Lisa.
LISA:    How cool.
```

Concluding that the tape fixating the crystals on the microscope plate confuses their perception, the teachers decide to take it away. When the crystal finally come into sharp view it is met by enthusiastic aesthetic responses such as `Oh, look!`, `Ooh!`, and even clapping (applauds). Children's attentive engagement is further indicated when one child saying to one of the teachers, `Yes. Did you see, Lisa?`, making sure the teacher has notices what the child has and considers worth attending others to (cf. Tomasello, 1999).

```
Child:  Lisa, it's so cool, this rock.
LISA:   Yes. It looked like...
Child:  It looks like a person.
Child:  It looks like a person with just one long arm.
LISA:   Oh yes.
Child:  Oh shit.
TINA:   Now we have a sensation here.
Child:  You can make a bigger picture too.
```

The visual pattern they see on the whiteboard triggers the children to express aesthetic judgments that they direct to and thus involve others in perceiving; one child makes one of the teachers attend to the pattern; `Lisa, it's so cool, this rock`. The teacher confirms the child's statement and proposes `yes, it looked like...`, which the children readily continue, saying that `it looks like a person. It looks like a person with just one long arm`. One child continues the other child's simile. A collaborative sense-making practice evolves. `O shit`, as exclaimed by one of the children, is actually said in English; this is a common expression in the Swedish young with a positive sense, used in a similar way as 'cool' (cf. Jakobson & Wickman, 2007).

```
LISA:   We can actually do it like this, you know, and I'm
        going to change that a little. Yes, I'm going to take
        that slightly bigger one there.
JOHN:   It should actually be kept flat... [inaudible].
TINA:   If we get more light on it, does that work? So that
        we can get that glittery appearance.
JOHN:   We can zoom in even more.
TINA:   I think I'll zoom in one more click.
JOHN:   [inaudible]
TINA:   What does it look like?
Child:  Chickens.
Child:  A cliff.
TINA:   A cliff, yes.
Child:  I think it looks like a...
Child:  Or grass.
```

```
TINA:    Ah, but look here!
Child:   Dragon-mushroom.
TINA:    There's the glitter. Look here.
Child:   It looks like silver!
Child:   It looks like a dog. With a tail.
Child:   No.
Child:   But up there.
Child:   Look.
Child:   Ah, but can't you see it looks like a…it looks like
         when you make…
```

After some initial deictic talk (`that`, `that`, `there`), one of the teachers mentions that the attempt is to `get that glittery appearance`. Again zooming in, a teacher asks the children, `what does it look like?` One child suggests `chickens`, while another says `a cliff`. The latter simile, in contrast to the first one, as it appears in terms of her response, makes sense to the teacher (`a cliff, yes`). Other suggestions are made by children, for example, saying that it looks like `grass`. However, this suggestion is not responded to by the teacher. Instead, she exclaims, `ah, but look here!` Before verbalizing what she wants the children to see, a child cuts in with a neologism, `dragon-mushroom`. The teacher clarifies that `there's the glitter. Look here.` She directs the children's attention to what is on the whiteboard in certain terms. One child confirms this perception by using a related expression, `it looks like silver!` while other children negotiate but do not at this point reach an agreement on how to see the visual pattern. While talking about what something *looks like*, rather than what *it is* (the latter they have already decided beforehand), it is not arbitrary in terms of what to see the phenomenon. Certain similes make sense to others (the teachers and other children) and some do not. Even if the similes used in these conversations may be unconventional, learning to perceive phenomena in terms of something familiar that makes sense also to others, making possible the coordination of perspectives and sense is important. Such coordination work of how to perceive something in certain metaphorical terms is also done by scientists in laboratories trying to make sense of experimental observations, as studied by Ochs and colleagues (1996).

```
Child:                  Glitter.

[The children talk a little while somebody tries to adjust the
focus]

TINA:                   This is as big as we can make it. Now
                        we're as close as we can get.
LISA:                   Yes.
Child:                  How far away can we get.
TINA:                   We have looked at the very top, now
                        we are going to look right down at
                        the bottom. Now we do this.
TINA:                   Now it's starting to gleam, there.
                        Look.
Child:                  It looks like silver.
TINA:                   Yes, it's glittering now. Yes.
```

[The children point and look]

Child: What is it actually?
All the adults answer: It's the salt.
TINA: That we had in jar. This is the salt. Magnified.

One child uses the description, glitter, as previously introduced by the teacher. When the image once again comes into focus, the previously introduced (by child and teacher, respectively) descriptions recur, it looks like silver and yes, it's glittering now, and these perceptions are thus coordinated. At this point something very interesting occurs; the children point at the image on the whiteboard and one of them asks, what is it actually? This question, what something really *is*, is very different to the conversation thus far which has revolved around the issue of what something *is like* (how its appearance could be described). Interestingly, this question comes from the children, not from the teacher. The teachers respond, It's the salt. That we had in jar. This is the salt. Magnified. The conversation continues:

Child: [inaudible]
TINA: No, but it is magnified. How many times magnification is that now?
LISA: Uh…forty times.
TINA: Forty times.
MARIE: Next time will do sixty times. Because then you can't see what it is.
Child: No.
Child: If you go further away, you can see it much better.

[The children and adults chatter]

Child: Yeah, there it is.
MARIE: It's starting. Glittering a little.
Child: That looks like silver.

[Children and adults continue to chatter]

TINA: Shall we see if we have done it.
MARIE: I think we have to zoom out a little to make it visible. You can't see it as clearly with these…is that forty?
JOHN: Yes.

Unfortunately, the child's talk is not discernible, but judging from the teacher's response, no, but it is magnified, suggests that the child has expressed some doubt. After having zoomed in and out for some time and talked about what the image looks like, all children do no longer seem to be clear about what it is they look at (see the previous excerpt). The child saying that if you go further away, you can see it much better indicates an understanding of the principle that being able to see clearly through the microscope implies a 'trade-off' between magnifying and getting too close.

MARIE: Let's try twenty and have a look.
TINA: Yes, we'll try twenty too. I don't think we'll see anything. Because it becomes so blurred and so…

[The focus is improved]

Child: Wow!
MARIE: Wow!
TINA: Yes it was, it's like this! This is what it should look like.
Child: Cool.
Child: Take a picture!
Child: Take a picture!
TINA: We can take a picture there.
LISA: Do you know how to do it Marie?
MARIE: Oh yes.
LISA: Yes.
TINA: Can you imagine that those white things look like this. The salt crystals in our jar. That they look like this magnified. That's fantastic, isn't it. It's almost unbelievable.
MARIE: Yes, it's like they are cerise...
TINA: It's fun to see it like this.
Children: Yes.
TINA: I think so.

Several aesthetic exclamations are heard among the children and teachers, wow!, wow!, cool, and the children in their enthusiasm tell the teacher to take a photograph of what they see. At this point, when they children are greatly attentive to the image on the whiteboard, one of the teacher makes explicit the connection between what they see and what it is, Can you imagine that those white things look like this. The salt crystals in our jar. That they look like this magnified. That's fantastic, isn't it. It's almost unbelievable. Some more aesthetic appreciation is heard, it's fun to see it like this.

JOHN: It must be fairly flat.
MARIE: Oh, yes.
TINA: Yes, that's right, so that it fits in the picture. There!
MARIE: Wow.

[Picture of crystal appears on the screen again]

Child: There!
Child: Ooh!
Child: What's that long thing there?
TINA: It might be something...
JOHN: It might be a strand of hair or something lying there.
MARIE: Yes, something that ended up there.
Child: Or it might be, or it might be, it's this, what's it called...
JOHN: Or the threads, perhaps.
Child: Yes, the threads, the threads!
JOHN: Yes, the threads, I should think.
TINA: It might be the threads it is climbing on.
Child: That's brilliant!
TINA: Oh, look!

```
Child:  Take it, take it, take it.
Child:  Take it, it's cool.
Child:  That.
Child:  That was cool.
```

The by now familiar aesthetic expressions and deictic references recur. More interestingly, at this point, is the conversation triggered by one of the children asking, `What's that long thing there?` As evident from the teacher's response, it is not clear what it is they see, `it might be something`... Another teacher ponders whether it `might be a strand of hair or something lying there`. Some vague attempts are made to suggest what they see, and a child starting to say something but does not appear to find the word looked for is continued by one of the teachers suggesting, `or the threads, perhaps`, that is the thread that the salt crystals are attached to. The child responds with emphasis, exclaiming `yes, the threads, the threads!` This suggestion is confirmed, `yes, the threads, I should think. It might be the threads it is climbing on`. The expression (Swedish: `ball`), as used by the children here is a positive expression, similar to 'cool', which has been used as translation here. After some further looking under the microscope, the teachers and children return to their classroom and the activity is concluded.

9.5 Discussion: What a Phenomenon Looks Like and What It Is

In this chapter we have followed an extensive sequence of teachers and children talking about and under a microscope, projected to a whiteboard, making sense of what they see and how this could be described. Particularly, the analysis of the data has emphasized the important distinction and relationship between what something *looks like* and what it *is*, as seen from a certain perspective, domain of knowing. Throughout the sequence followed, the issue of aesthetic judgments (Jakobson & Wickman, 2007) has also kept reappearing. In the final parts of this chapter we will discuss these matters and what they imply for children's science learning. A critical issue when learning science as well as other representational forms of knowing, are to be able to discern the object or phenomenon represented (modeled, illustrated) from the mode of representing it. This issue comes up for negotiation between the teachers and the children on several occasions throughout the followed lesson. What *is it* that they see through the microscope as magnified on the whiteboard, is it tape, dust, a thread or is it the salt crystals as such?

Another important observation was that the children themselves used markers, clarifying that they have some idea about the phenomenon discussed but that they cannot yet clarify more precisely *what it is*. These children are thus on the way in their knowledge development, they *know what it is like* and that this simile is somewhat correct but not quite what it is (cf. Pramling, 2006). As seen in several chapters of this book, a recurring speech pattern could be described as going from deictics to verbalization of meaning. Deictic references are those communicative actions that point to something, either physically with one's fingers or verbally through words such as 'that', 'there', 'it', and 'then' (Ivarsson, 2003; Rommetveit, 1968). Verbalising deictic references clarifies what these refer to and how they should be represented in speech (categorized, labeled) in this particular activity. To do so is also important in order to coordinate perspectives between different communicative partners (e.g., a teacher and a child). This clarifying follow-up action is often done by teachers. However, as seen in the present chapter, there are also occasions when the children do so, that is, a child verbalizing a teacher's deictic reference. The teachers in this lesson communicatively frames (Goffman, 1974) the activity in ways that are subsequently picked up and aligned with by the children. One example of this is in the beginning of the lesson when one of the teachers uses an active metaphor when speaking about the formation of the salt crystals and a child reuses this way of speaking, in effect constructing a kind of proto-narrative (cf. Pramling & Ødegaard, 2011). Another example is how the teachers introduce aesthetic judgments into the activity, something the children also take on and use. As earlier pointed out by Jakobson and Wickman (2007) and Wickman (2006), aesthetic judgments such as 'beautiful' etc. are common in science learning.

While the participants in the followed activity frequently described what they perceived what they saw on the white board *as*, that is what is was *like*, they did less frequently engage in conversation about *what it is*. Letting children suggest what something looks like and to suggest such similes and metaphors as a teacher facilitates the children's engagement through allowing them to use what they already know and are familiar with as resources for making sense of and communicating to others something that is less well known and what they are only beginning to familiarize themselves with. However, in order to build upon these resources, a more knowledgeable interlocutor (e.g., a teacher) needs to relate these to a more established or conventional language (discourse) for speaking about the phenomenon. Without such relational work, and if only confirming the child's suggestions of what something looks like, the child will not be supported in appropriating new cultural tools useful for making sense of and communicating about nature. Alternatively, if only introducing what something is, as understood conventionally within a science discourse, the novel will not make sense to the child in being unrelated to what he or she already knows. If taking a Vygotskian (1987) perspective on conceptual development, there is an inherent dynamic and necessary tension between everyday and scientific concepts in the child's development (see Chap. 1). Seen in terms of the distinction made in this chapter, what something *is like* needs to be in speech negotiated in relation to what it in a certain discourse *is*.

The communicative frame established by the teachers leave much space for children's playful similes in suggesting what they perceive the phenomenon observed looks like, and they readily engage in this description with great joy (laughing, applauding). However, the teachers, in addition to encouraging the children to more freely say what they think it looks like, direct their attention to what is on the board in certain terms (e.g., `glittering`) which is subsequently taken up by the children in describing what they see. Two or more people sharing attention on something third is fundamental not to learning, which is a much wider concept, but to what may be referred to as an education (Pramling & Pramling Samuelsson, 2010; see also Chap. 11). To establish in speech how to mediate or represent what is observed is one way of coordinating perspectives with this end. We now turn to graphical representations in science and what this means for scientific thinking of very young children.

References

Bruner, J. S. (2006). Narrative and paradigmatic modes of thought. In *In search of pedagogy, volume II: The selected works of Jerome S. Bruner* (pp. 116–128). New York: Routledge.

Fleck, L. (1979). *Genesis and development of a scientific fact* (F. Bradley & T. J. Trenn, Trans.). Chicago, IL: University of Chicago Press. (Original work published 1935)

Goffman, E. (1974). *Frame analysis: An essay on the organization of experience*. New York: Harper & Row.

Heath, S. B. (1996). *Ways with words: Language, life, and work in communities and classrooms* [with a new Epilogue]. New York: Cambridge University Press. (Original work published 1983)

Ivarsson, J. (2003). Kids in zen: Computer supported learning environments and illusory intersubjectivity. *Education, Communication & Information, 3*(3), 383–402.

Jakobson, B., & Wickman, P.-O. (2007). Transformation through language use: Children's spontaneous metaphors in elementary school science. *Science & Education, 16*, 267–289.

Ochs, E., Gonzales, P., & Jacoby, S. (1996). "When I come down I'm in the domain state": Grammar and graphic representation in the interpretive activity of physicists. In E. Ochs, E. A. Schegloff, & S. A. Thompson (Eds.), *Interaction and grammar* (pp. 328–369). Cambridge, UK: Cambridge University Press.

Pramling, N. (2006). 'The clouds are alive because they fly in the air as if they were birds': A re-analysis of what children say and mean in clinical interviews in the work of Jean Piaget. *European Journal of Psychology of Education, 21*(4), 453–466.

Pramling, N., & Ødegaard, E. E. (2011). Learning to narrate: Appropriating a cultural mould for sense-making and communication. In N. Pramling & I. Pramling Samuelsson (Eds.), *Educational encounters: Nordic studies in early childhood didactics* (pp. 15–35). Dordrecht, The Netherlands: Springer.

Pramling, N., & Pramling Samuelsson, I. (2010). Evolving activities and semiotic mediation in teacher-child interaction around simple objects. *Educational & Child Psychology, 27*(4), 22–30.

Pramling, N., & Pramling Samuelsson, I. (Eds.). (2011). *Educational encounters: Nordic studies in early childhood didactics*. Dordrecht, The Netherlands: Springer.

Pramling Samuelsson, I. (2006). Teaching and learning in preschool and the first years of elementary school in Sweden. In J. Einarsdottir & J. T. Wagner (Eds.), *Nordic childhoods and early education: Philosophy, research, policy, and practice in Denmark, Finland, Iceland, Norway, and Sweden* (pp. 101–131). Greenwich, CT: Information Age.

Rommetveit, R. (1968). *Words, meanings, and messages*. New York: Academic Press.

Tobin, J. J., Wu, D. Y. H., & Davidson, D. H. (1989). *Preschool in three cultures: Japan, China, and the United States*. New Haven, CT: Yale University Press.

Tomasello, M. (1999). *The cultural origins of human cognition*. Cambridge, MA: Harvard University Press.

Vygotsky, L. S. (1987). *The collected works of L. S. Vygotsky, Volume 1: Problems of general psychology, including the volume* Thinking and Speech (R. W. Rieber & A. S. Carton, Eds., N. Minick, Trans.). New York: Plenum.

Wertsch, J. V. (2007). Mediation. In H. Daniels, M. Cole, & J. V. Wertsch (Eds.), *The Cambridge companion to Vygotsky* (pp. 178–192). New York: Cambridge University Press.

Wickman, P.-O. (2006). *Aesthetic experience in science education: Learning and meaning-making as situated talk and action*. Mahwah, NJ: Lawrence Erlbaum.

Winner, E. (1988). *The point of words: Children's understanding of metaphor and irony*. Cambridge, MA: Harvard University Press.

Chapter 10
Learning to 'Read' and Produce Graphical Representations in Science

Niklas Pramling

Abstract In this chapter figurative (metaphorical and similar) speech is discussed in the context of the distinction between "as is" and "as if" statements. Representational practices and modeling phenomena are all examples of semiotic mediation. Since learning is understood as the appropriation of cultural tools that come to mediate the learner's engagement with the world, how teachers introduce children to, and support them in appropriating such tools, is the focus of attention of this chapter. It is argued that what others have experienced can be made into cultural tools. That is, these cultural tools are represented as artefacts, such as speech, writing, images, drawings, pictures and recordings. Throughout history, human experience and knowing have collectively accumulated in the form of cultural tools and artefacts. Empirical data on the topic of teaching about the human body taken from a preschool is presented in order to examine how teachers allow or even encourage children to 'take' or use representation tools (e.g. drawing) and metaphorical speech during conceptual play.

Keywords Human body • Drawing • Metaphorical speech • Representations

10.1 Introduction

Figurative (metaphorical and similar) speech, the distinction between as is and as if, representational practices, and modeling phenomena are all examples of semiotic mediation (Wells, 2007; Wertsch, 2007), that is, the fact that we do not have immediate access to the world but through the cultural tools that we have appropriated or are in the midst of appropriating. As was introduced in Chapter 1, Mediation was a revolutionary idea in Vygotsky (1987, 1997, 1998) and basic to a cultural-historical perspective on human learning and development. How this issue of mediation comes into play in teacher-child interaction in early childhood education, what it entails for children's development and teachers' role in this development are the themes of this chapter, focusing on representational practices in early childhood science learning. Since learning is understood as the appropriation of cultural tools that come to mediate the learner's engagement with the world, how teachers introduce children to, and support them in appropriating, such tools are key interests of this chapter.

M. Fleer, N. Pramling, *A Cultural-Historical Study of Children Learning Science*, Cultural Studies of Science Education 11, DOI 10.1007/978-94-017-9370-4_10

One important feature of a cultural-historical perspective on human learning and development is the notion of tools (Daniels, 2005; Kozulin, 1998; Säljö, 2005; Tomasello, 1999; Vygotsky, 1978). Throughout history people have developed tools that allow us to externalise knowing, make it public and give it permanence over situations and generations. These tools come to mediate learner's engagement in the activities where they are used. We learn about the world in a 'roundabout' way, to use Vygotsky's metaphor (Fleer, 2009; Vygotsky, 1991). This means, among other things, that we do not each and every one have to experience something in order to know something. What others have experienced can be made into cultural tools, that is, be represented in artefacts such as speech, writing, images, drawings, pictures and recordings. Throughout history, human experience and knowing are collectively accumulated in the form of cultural tools, artefacts. This has important implications for educational practices. Learning to great extent becomes a question of appropriating (taking over, learning to produce and 'read') various kinds of representations. Some examples would be texts (literature), diagrams and models in science education. A particular kind of representation is metaphorical speech, in saying that one thing is another thing that it cannot in a strict sense be. This kind of speech shares with a visual representation (e.g., a schematic drawing of the inside of the human body) that it is 'tricky'. The learner/reader must learn how to 'take' the representation (i.e., in what sense is what is spoken about [like] this representation). Clarifying this matter could be seen as a re-contextualisation (van Oers, 1998), of relating previous experience and language to something novel. Where the representation or metaphorical speech 'takes' the child will likely be contingent upon how it is responded to and managed by the teacher. Does the teacher, for example, allow (or even encourage) the child to 'take' the representation/metaphorical speech in one or the other direction (conceptual play)?

In order to investigate and shed some light on these issues, in this chapter, we will analyze empirical data from early childhood education (a Swedish preschool). The empirical data for the present analysis comes from a theme on the human body, conducted over three consecutive arranged learning situations in a preschool with children. A small group of children, varying between the three occasions but at most six children and their teacher participate in the activities. The activities take place once a week for three weeks in a row. During these occasions, they look at visual representations in books, make drawings and speak about these and the phenomena they represent. Hence, how the teacher and children manage matters of representation, that is semiotic mediation, is analyzed in this chapter.

10.2 Previous Research on Children's Drawings in Science Class

There is quite an extensive research literature on children's understanding of the human body and its organs (e.g., Carey, 1985; Cuthbert, 2000; Guichard, 1995; Óskarsdóttir, 2006; Tunnicliffe & Reiss, 1999). This literature is often based on constructivist theory and reports what is referred to as children's 'misconceptions' and at what ages children develop certain insights. However, we will not review this literature here, for several reasons (for overviews, see STCSE database). First, taking a very different theoretical perspective, what children say and do are interpreted differently (Pramling, 2006b) and the kind of data that is used also tend to differ. In addition, from our theoretical position, it is highly problematic to conceive of children's reasoning as being 'misconceptions'. Rather, they make sense of the task as they perceive it and their ways of reasoning is highly contingent on the nature of producing data, including how questions and tasks are communicatively framed (Aronsson & Hundeide, 2002; Goffman, 1974; Säljö & Wyndhamn, 1993). All these matters are elaborated upon in the present book (see particularly Chap. 7), and in this chapter. Finally, the issue we thematise in this chapter is not how children understand the human body and how that understanding changes with age, but theoretical issues having to do with representation and the coordination of perspectives between a teacher and a child. We will therefore primarily refer to a different body of literature. Still, we will here overview studies where children have been asked to make drawings of the human body (and then talk about these drawings).

In their study of drawing during science activity in primary school, Hayes, Symington, and Martin (1994) suggest that there are several reasons for letting children make drawings during science activities. First, the common observation that (most) children like to draw is in itself a reason to let them do it, since, Hayes et al. write, "The enjoyment they derive is likely to be important in providing the motivation for engaging in similar activities in the future" (p. 265). While being sympathetic to the point made, we may add that "similar activities" from the child's point of view may be drawing rather than science activities. Hayes et al. add another reason for including drawing activities in science class. They divide this reason in two groups. The first, referred to as 'objective purposes' denotes the expectation that the activity will result in children developing certain abilities, such as observing or understanding phenomena investigated. The second reason for including drawing, referred to as 'process purposes', denotes the idea that the activity will develop in the children other skills such as communication skills as well as keeping the teacher informed about how the children think about phenomena. While children's drawings can be informative as to how they understand something, there are many additional issues to keep in mind. For example, in contrast to earlier psychological theorizing on children's drawings, these are no longer considered 'windows into the child's thinking or understanding' in any clear-cut manner (Bendroth Karlsson, 1996). There are several reasons for this, including the fact that drawing for children to a large extent is a social activity where they feed off each other (Änggård, 2005) and that representations – whether in the form of drawing, writing or other modality (Kress, 1997, 2003) – never stand in a simple one-to-one correspondence with what they refer to (Pramling, 2006a). In fact, the latter issue of the dynamic tension and potential developmental relationship between representation and its reference lies at the heart of what will be analysed as the participants' concern in this chapter.

Osbourne, Wadsworth, and Black (1992), among many different data sets, used children's drawings to investigate their understanding. One finding was that the youngest children (who were 5–6-years old) drew organs that they could easily discern such as the heart (through its beats) and the bones. As found in the study,

> the predominant organs named by all children were the heart, bones, stomach and brain. The study also revealed that many children were not aware of the correct size or the location of the organs which is probably because the internal organs are not visible or touchable and therefore it is difficult for the child to develop knowledge of their size and correct location. Organs that are not part of everyday language like kidneys, liver, intestines and even lungs were usually excluded by the children although most of the children knew that we need air to live but very few were able to locate the lungs on a drawing of the human body. (Osbourne et al., 1992, p. 37)

This finding also, from our point of view, implies the importance of attending to issues such as representation in speech, including how tasks are given, questions are posed, terms explained, and perspectives are coordinated.

In a study of almost 600 pupils from 11 different countries, Reiss et al. (2002) gave pupils a blank piece of paper and asked them to "draw what they thought was inside themselves" (p. 58). The pupils were 7 and 15 years old, respectively. The overarching question for the study was whether the pupils' knowledge of their insides is "dependent on their culture" (p. 58). The drawings were graded by the researchers according to a predetermined scale "where the criterion was anatomical accuracy" (p. 58). If taking a cultural-historical perspective, there is no neutral way of giving a task, and therefore we cannot presume that all the children intended to make anatomically accurate drawings. There are always the issues of how a task is given and taken, and it is well known that subtle differences in how tasks are given are of decisive importance to how people act in response to these (Aronsson & Hundeide, 2002; Donaldson, 1978; Hundeide, 1977). The task was given to the pupils in the following words: "I would like each of you to do a drawing of what you think is inside yourself" (p. 59). If a child was to say that he or she could not draw, the researchers in the various countries were instructed to say that he or she need "not to worry and that we are interested in what they think is inside themselves not in whether they can draw well" (p. 59). Two things are noteworthy with the last comment. First, that the drawings are seen as more or less unproblematic pictures of children's thinking and second, that despite this instruction, the drawings were subsequently in fact analysed in terms of "anatomical accuracy", that is, how well they represented the inside of the body. However, the oversimplified stance taken towards the drawings are noted by the researchers themselves. For example, they reproduce a drawing made by a Chinese student containing not only labels on the drawing such as 'cell', 'blood' and 'heart' but also 'future' and 'money'. As for the cultural dependency of the children's understanding, only minor examples are given, and no systematic analysis and result in this regard is reported in the study.

It should be remembered that the children of the present chapter are far younger than the children covered in the research studies here briefly reviewed. Also, the situation of drawing was very different to the task-like nature of previous studies. What we follow and analyse is the unfolding nature of how a teacher and children communicate about and through the drawings the children make, particularly regarding issues concerning how to produce and 'read' representations.

10.3 Empirical Study

10.3.1 Introducing the Theme

The first event is introduced by the teacher and the children sitting in a circle on the floor. The teacher reminds the children of something they had found in a cupboard another day:

```
Teacher:     We found small eggs, in a bag, so we mixed them
             together with a little water. So we put the eggs
             in the water. [---]
Teacher:     Do you remember what type of eggs they were?
Children:    No
Teacher:     What were they going to grow into?
One child:   Shrimps.
Teacher:     Shrimps, yes. You can hardly see them, they are
             so small. [Takes out a glass bowl and shows the
             children]. You can just see that it is a little
             dusty in there. Can you see that?

[The children look carefully into the bowl]

Teacher:     How small are they?
One child:   I can't see the eggs.
Teacher:     No, they're so small. You can look one at a time.
             Do you see? That it is, it's a little cloudy,
             like tiny grains of sand.

[The children get to look in the bowl, one by one]

August:      I saw the eggs.
Richard:     I can see loads of brown peas.
Teacher:     Yes, and I have actually put some of these eggs
             in the microscope over there. Although they will
             be big on the screen.
```

Supporting the children in remembering what event she refers to, the children recall that the eggs would become shrimps. Interesting to note in this initial excerpt from the activity is also how the teacher and children verbalize what they see. Looking at the eggs, the teacher suggests that the eggs are hardly perceivable, `You can just see that it is a little dusty in there.` Words like `just a little` and `it is, it's a little cloudy, like tiny grains of sand` are forms of markers that hint at the difficulty of seeing (visually as well as conceiving) something and that therefore a simile is used. In response to this suggestion, Richard says that `I can see loads of brown peas`, that is, he describes what he sees in familiar terms from another domain. The children then get to look at the eggs under the microscope.

In the evolving conversation between the teacher and the child, babies are introduced. Speaking about a child's mother having a large belly this transition in the conversation is made.

Teacher: A little baby. And we will see what happens with the little thing, that's inside there. [...] may stand here. But I also thought that we should talk about what we look like inside. Because you are growing all the time, aren't you? Do you think that you will grow bigger than you are now?
Eva: I'm really big, this big [stretches up her hands].
Teacher: Do you think you are going to get bigger?
Eva: [Nods]
Teacher: We will look at what we have inside our body.
Dennis: Yes.

Hence, during the conversation around the eggs and the microscope, the talk goes from shrimp eggs to human fetuses to how humans look on the inside. This is a kind of analogical reasoning that leads to the theme to be worked on: the (inside of the) human body.

10.3.2 Representing the Human Body in a Drawing

The teacher now takes out a large sheet of paper and crayons and says that they shall draw a body, Just like you look. Then you have to think, how do we look? What do we have up here? To this several children reply, Head. The teacher confirms the children's suggestions and continuing drawing the outline of a human body on the paper, asks the children, And what do we have below the head? and, What's this (showing on her body)? The children respond, shoulders, throat, stomach, chest, arms, hand, bottoms, legs...

Teacher: Legs go there, yes. [draws] Now we are looking at this body from in front, so that we can't actually see the bottom. We can do this [draws a little around by the hips] so that we know that the bottom is behind here. Then we have the legs.

This explication from the teacher introduces an important issue for our present concerns, that is, the issue of representation and how to represent what one knows is there but cannot be seen from a certain point of view. Another important issue that comes into play when representing something is aesthetic preferences. When drawing the eyes of the person on the paper, Polly suggests they be pink. Drawings in preschool often take on a kind of hybrid form where issues of representing something ('accurately'), on the one hand, are intertwined with the issue of drawing something nice, beautiful or expressive, on the other (cf. Bendroth Karlsson, 1996; Kress, 1997; Pramling & Pramling Samuelsson, 2011). The issue of the aesthetics of the drawing of the human body will recur throughout the activities. Central to representing something is of course also what the representation is a representation of. When coming to drawing the face, the teacher asks, where is the chin on the man? This question prompts Dennis to respond, it's not a man, `it's a girl`. This suggestion may be due to the fact that earlier in the activity, they had spoken about and drawn breasts on the figure. One of many examples of the issues of aesthetics and referent of the representation coming into play is the following exchange when one child says `He is fine` and the teacher responds, `Really fine, or is it a she? One doesn't really know huh?`

Having drawn the outline (contours) of a body, the theme of the *inside* of the human body is introduced by the teacher in the following way:

```
Teacher:    Now I have a question, actually, before we finish.
            What is behind our eyes, cheeks and nose [points
            to her own face]? What do we have inside?
Richard:    Blood!
Teacher:    There is blood inside, yes. I'm going to write up
            what you say now. We have blood, what else do we
            have inside our head?
One child:  I know! Our brain!
Teacher:    Our brain, yes. What do we use our brain for?
August:     To suck up the blood.
Teacher:    [Writes on the paper] Blood, we'll write that
            here. And brain…to suck up the blood, we'll write
            that there.
August:     Yes.
```

Another mode of representing is thus introduces along with the new focus on the inside of the human body, through text (words and instructive comments). This means that the representation becomes more complex. In difference to outlining a human body, for example, following with a pen the extension of the fingers on one's hands, words do not stand in such a simple and iconic relationship to what is represented.

Connecting to the issue of blood, the teacher directs the children's attention to a visible feature:

```
Teacher:  Have you seen this? Look here! Look at me, at my
          arms [shows blood vessels on her hands and arms].
          Do you know what this is? This blue thing. [Polly
          continues to try to say something about it sucking
          blood, but Helen continues to point and ask. Polly
          becomes quiet and shakes her head.] It's actually a
          little tube, you could say, so that the blood can
          travel inside. This blood is on its way back to my
          heart. What is the blood doing here?
Polly:    And Cator actually has a heart.
Teacher:  Yes, he has. That's in the Brothers Lionheart.
```

In explaining what it is they see, the teacher uses a metaphor and a marker, `It's actually a little tube, you could say`. Words like `you could say` and `actually`, as paradoxical as the latter may seem, are frequently used to signal that something is being spoken about in more familiar but not entirely correct words, that is, that the utterance is figurative rather than literal (Goatly, 1997). One of the children responds to the teacher's question by relating to a character in a fictional story. The children are now weary, after having worked on the human body for half an hour and the activity ends.

10.3.3 The Parts and Functions of the Human Body and How It Is and Can Be Represented in Drawings

When the group of children a week later (this time only three of the six children from the first event is present, and their teacher) meet in the 'nature group', one of the children, August, immediately connects to last week's topic, `There was also skeleton` (points at the outlined person on the large paper). The children suggest more features of the body, such as hair and nose and that you `blink` with the eyes. The teacher connects and expands:

```
Teacher:            Yes, we blink with our eyes. What else do
                    we do with our eyes?

[Music and talk in the background makes it difficult to hear]

Malin:              Have medicine in them.
Teacher:            What?
Malin:              Have medicine in them.
Teacher:            Yes, we can have medicine in our eyes if
                    they are poorly, quite right. But if we
                    close our eyes, can we see anything then?
Several children:   No.
Teacher:            And if we open our eyes, we do we do then?
Children:           See!
Teacher:            See, yes. We use our eyes to see with.
                    Don't we? To see what is happening. Then
                    we have these things over our eyes [points
                    to her own eyebrows].
```

The teacher thus introduces the issue of the functioning of different body parts, that is, what we do with the eyes. The teacher confirms the child's uptake, that we can take medication in the eyes, but then introduces another issue, `But if we close our eyes, can we see anything then?` Continuing along these lines, the conversation continues:

```
Teacher:   What do we do with our nose?
Malin:     Holes.
Richard:   Smell with.
Teacher:   We have it to smell with, yes. And we have little
           holes in our nose as well, yes, we have two little
           holes in it as well. Mm. And then, can we do anything
           else with our nose. Do you know what it is called
           when we do this? [Shows how she is breathing in]
Richard:   Bogies!
Teacher:   Yes, you have some bogies in your nose, yes, which
           makes it a little for you difficult to do it. We
           breathe with our nose. And we also breathe with our
           mouth. What happens when we breathe? Do you know?
August:    Are we going to do the skeleton?
Teacher:   Yes, we are going to do the skeleton. What happens
           when we breathe? We take in a lot of air, like this
           [breathes in deeply]. What happens then? Where
           does the air go?
```

Richard: Away!
Teacher: It disappears somewhere down here in the body, doesn't it. If you hold here. [Shows on August] And you breathe in. It becomes so big there. And then you breathe out. There is a book here. You can see what it looks like inside your body. This is an old book I found. [leafs through the book]. Then we will see what happens. Here we have what August was talking about [points to the picture]
August: The skeleton.
Teacher: Here is the skeleton, yes. What is the skeleton made of, then? Do you know? Is it hard or soft?
Children: Hard.
Teacher: It is hard, yes.

The children's responses are relevant and the teacher confirms them, but she also redirects their attention towards other aspects of the nose (or earlier the eyes) from smelling to breathing. After giving a first explanation of breathing, the teacher picks up a book with drawn illustrations of the inside of the human body, including the skeleton which seems to particularly attract August's interest. The teacher and the children look at the illustrations and speak about what they see:

Teacher: We have a lot of skeleton here, don't we. [Shows August] Throughout our body. Now we will see if we can find out about… what we were talking about before. Can you see that here is the head? And you said that there is a brain and that we suck up blood with it. And here is the brain, actually [points in the book]. Now the brain probably isn't blue. They have drawn it with lots of strange colours. Do you know, what colour do you think the brain is inside?
Malin: Blue.
Teacher: You think it is blue, OK. What do you think, August?
August: Green.
Teacher: Green, you think? What colour do you think the brain is Richard?
Richard: Red!
Teacher: Red, you think? Why do you think it is that colour?. Why do you think it is green [turns to August]?
August: I think because it is green here [points to the picture].
Teacher: Mm, on the picture. This picture is fooling us, actually, because the brain is not really green or blue. Why do you think the brain is red, then? [Asks Richard]
Richard: It is inside the body.
Teacher: It is inside the body? Mm.
Teacher: It's probably a little more reddish-brown, yes. And that is actually true, because it is actually. You also said that there is blood in there. Didn't you? And that's also true.
Richard: And it gets, there are bumps as well.
Teacher: Yes, there are bumps. What is a bump?

Richard:	[Points to his head and says something inaudible]
Teacher:	It's actually if you knock yourself, you get blood inside here. Then it becomes a bump. Then it turns blue.

This sequence is of particular interest to the issue of representation. The relationship between the brain and blood as introduced earlier during the first event is now returned to. In addition, as triggered by the illustration in the book, the issue of what color the brain is, has been raised. In the drawing it is blue. However, as the teacher cautions the children, Now the brain probably isn't blue. They have drawn it with lots of strange colours. Still, when asked if they know what color the brain is, the children attend to the colours of the drawing. The teacher points out that the image may not be adequate in this regard, This picture is fooling us, actually, because the brain is not really green or blue. While Malin and August have taken the colour of the drawing literally, as showing the colour of the brain, the third child, Richard instead suggests red. Asked to clarify why he believes the brain to be red, Richard in a somewhat shorthand way says, It [that which] is inside the body [is red]. Probably on the basis of experience of blood being red, Richard proposes that what (all that) is inside the body is red. In this way he disconnects the relationship between the representation (the drawing) and what it refers to and instead builds upon his experiences from elsewhere. The teacher confirms his observation in these terms, that is, in terms of blood being red. In this excerpt, the relationship between the representation and what is being referred to is thematised in conversation between the teacher and the children as not being of a simple corresponding, depicting, nature. Learning what to take as representing something and what is simply a feature of the representation as such, is an important lesson in science education.

After August having told about him bumping his head on a door knob, the teacher redirects the children's attention to the book she holds:

Teacher:	Yes, there it is [Shows the book] Yes, we can see everything here, that's a pity. It's like this, this is also there, inside here, inside your chest in there, there is something here [points to the heart]. Do you know what this is sitting in here?
Children:	No.
Teacher:	It's something that says this: "Donk-donk, donk-donk".
One child:	The heart!
Teacher:	It's the heart, yes. Does anyone want to paint the heart?
Several children:	Yes!

Continuing to speak about what is inside the body, the teacher asks what is behind the ribs. When the children respond that they do not know, the teacher uses a metaphorical utterance, It's something that says this: "Donk-donk, donk-donk", that is something (the heart) 'says' what she illustrates through onomatopoeia (i.e., an expression mimicking the sound of something).

With this assistance, the children in unison exclaim, `the heart!` Since all the children want to draw the heart on the paper they begun last week, the teacher uses a rhyme to arrive at August as the one who gets to draw it this time:

Teacher: Eeny, meeny, miny, moe, etc. … Ok, August gets to paint the heart. Does anyone know what colour, or, what colour do you want to use? [asks August]. Blue?
August: No [puts back a blue chalk], the heart is a red colour.
Teacher: The heart is red. Wait a minute [stops August]. Where does the heart go on this man or woman, this person? Where does the heart go?
August: Here [points].
Teacher: The heart goes here, yes.

[August draws the heart in the right place]

Teacher: Are you going to fill it in as well?
August: Yes. Blood [fills in with red]. That's what the blood should look like.
Teacher: Is there blood in the heart? What does the blood do in the heart then?
August: It will be sucked down to the tube.
Teacher: To the tube, which tube? [holds out the book with the picture and August points].
August: See, there are tubes!
Teacher: Exactly, that's quite right. And what do the tubes do with the blood, then?
August: They suck out.
Teacher: They suck out the blood? It's actually, you can see it here [shows on her own wrist], you can see them, small tubes. The blood comes here, and this blood is going back, back. You also have them if you look. If you look here. [Points to the children's wrists.] Look here. Look, here you have small tubes. It's back. Do you have them as well, Richard, do you have any tubes on here?
Richard: Yes. [Lifts up his foot]. I have them on my foot, too.
Teacher: Do you have tubes on your foot too, where then?
Richard: There. [Holds out his foot and points to a toe]
August: I've been bitten by a mosquito.
Richard: Yeah, a bit of blood.
Teacher: A little blood [inaudible] and the heart, as you said, it sucks out the blood in the body. And when it does this, donk-donk [shows with her hand], it is actually pumping out the blood.

The teacher's initial question here implies the hybrid nature of drawings as common in preschool practices, that is between depicting something in an 'accurate' manner and/or drawing as one likes (e.g., what colours one finds beautiful), `Does anyone know what colour, or, what colour do you want to use?` August chooses red, since `the heart is a red colour.` The undecided issue from the first event regarding the sex of the depicted person comes into play when the teacher asks August where the heart should be drawn, `Where does the heart go on this man or woman, this person?` August fills out the contours of the heart while speaking (to himself), `That's what the blood should look like.` Hence, he motivates his choice of colour in terms of accuracy. The teacher connects to August's claim and extends the discussion, `Is there blood in the heart? What does the blood do in the heart then?` Working on the human body is not simply about depicting the parts of the body, the teacher recurrently directs the children's attention to the nature and functioning of these body parts. Hence, her agenda as it becomes evident in her questions and responses, does not simply aim at the children learning a list of names, but instead get an understanding of the human body and how it works. August is responsive to the teacher's questions and suggests, `It will be sucked down to the tube.` The metaphor of `tube` was introduced when looking at veins in the first event (see above). In talking with the child, the teacher uses the drawing of the book to coordinate the metaphor and its referent. The children and teacher look for veins (as evidence of the tubes) on their bodies, and notice additional examples. Finally, the teacher summarises the discussion and once again uses the gestural metaphor (showing with her hand the pumping of the heart) and its accompanying onomatopoetic expression, `donk-donk`.

Moving on from the heart, the stomach comes up for discussion:

```
Teacher:  Where is your stomach? If this is the heart [points
          to the drawing].
Richard:  Here. [Points to himself]
Teacher:  [Points to the drawing] There, yes. That's where
          the stomach is. And what is in the stomach?
Richard:  A little big hole. And food.
Teacher:  Shall we see? [takes out the book] This is the stom-
          ach [points in the book]. This is also in the wrong
          colours. We don't have any colours like this inside
          the body, actually. Purple and green and so on, but
          I think they have tried to highlight it. This, do
          you know what this is called? [Points to the intes-
          tines] What does this look like, almost? What does
          it look like, do you think?
Richard:  [Inaudible] …like a toothbrush.
Teacher:  Like toothbrushes?
Richard:  …like toothpaste.
Teacher:  Like toothpaste, yes.

[---]

Teacher:  Shall I tell you what it is called?
August:   Mm.
Teacher:  These are actually called your intestines, these
          ones. And this is the stomach, and this is where the
          food goes.
August:   In the st…
Teacher:  In the stomach. It's like a little bag that the food
          travels down into here.
```

Once again the issue of the arbitrariness of the colours of the organs in the book is commented by the teacher. In an interesting turn of phrase, the teacher pointing at the intestines on the image asks, `This, do you know what this is called?` and, `What does this look like, almost? What does it look like, do you think?` Hence, leaving aside for the moment the issue of how this body part is conventionally labeled, she opens up for the children to use their experiences and knowledge in making sense of what they see, what they think it looks like (cf. Chap. 9). Richard responds by suggesting, at first, `like a toothbrush`. However, this simile, expressed in this way, does not make sense to the teacher, as evident in her response, `Like toothbrushes?` Following up on how his utterance was taken by the teacher, Richard now rephrases what he means, `like toothpaste`. Phrased in this way, the simile makes sense to the teacher, responding, `Like toothpaste, yes`. Finally, the teacher introduces the conventional term, `intestines` and says a few words about its relation to another body part, `stomach` and `this is where the food goes`. With a new term and an explanatory simile the teacher responds to August's `In the st…, In the stomach. It's like a little bag that the food travels down into here.`

Continuing talking about the intestines and the stomach, the teacher asks if the children remember what it was called:

```
Teacher:  And after the stomach, when the food has been there,
          it carries on [points in the book]. Do you remember
          what this is called, then?
Malin:    Muscles.
Teacher:  Yes, they are muscles, yes these are a type of mus-
          cles, because they pump around, but they are called
          intestines.
Richard:  Yes, it's [inaudible] that I have.
Teacher:  Yes. [Points in the book] And this is actually the
          large intestine and this is the small intestine.
          The large intestine and the small intestine.
Richard:  [Unclear] the same as…

[---]

August:   D'you know what, he did a P.
Richard:  Yes, the same as Patrick.
Teacher:  Mm. Yes, a P, the stomach looks almost like a P
          there. There are the intestines. Do you want to fill
          them in too?
```

While clarifying that this term is not quite the expected or conventional one, the teacher still supports Malin's suggestion by motivating how `yes these are a type of muscles, because they pump around, but they are called intestines`. Drawing the intestines, August points out that this looks like the letter 'P'. The similes used by the children indicate their experiences. In this case, seeing in terms of a cultural symbol (the letter P), contingent on him growing up in a literate culture where children tend to pick up (notice, discern) this communicative and representational system well before they receive formal schooling in reading and writing. Metabolism is now thematised in the talk:

Teacher:	What happens to the food after it has been in the intestines, do you know?
One of the children:	Purple.
Teacher:	When the food first goes through the stomach, and when what is left over goes through all the intestines and when it comes out there, what happens then, do you think? Where does it go then?
August:	Down to the legs.
Teacher:	The blood actually absorbs the food here, taking what it wants. But what about what the body doesn't want?

In explaining what happens to the food, an animistic form of reasoning is used, as quite common in these kinds of explanations (Thulin & Pramling, 2009). In this case, the blood is said to `actually absorbs the food here, taking what it wants. But what about what the body doesn't want?` The blood and the body are thus spoken about in terms of intentional agents that `want` something. This kind of speech recurs during the activity.

Having spoken about what happens to the food in the body, lungs and breathing are introduced and spoken about. Here something interesting concerning the issue of representation comes up in the talk:

Teacher:	But do you remember now what we said? Oh. When we breathe. There was something else here. The heart took care of the blood, August. But there was something else beside the heart, around here, can you see? [points to the lungs in the picture in the book] What are these? Do you know what these are called? When you breathe, where does the air go?

[Children and teacher whisper - inaudible]

Teacher:	They are called lungs!
Richard:	Lopopopo.
Teacher:	What a lot of words there are in the body, aren't there? It's the lungs are around the heart there.

Richard responds to the introduction of the new term, `lungs`, with a nonsense word that sounds somewhat similar in Swedish (`lungor, lopopopo`). Being sensitive to the children maybe finding all the terms introduced somewhat overwhelming, the teacher says that `What a lot of words there are in the body, aren't there?` In this way, the difference between the representation (in this case verbal terms) and what is referred to (bodily organs) is collapsed into one and the same, the body containing words (cf. Pramling & Säljö, 2007, for an analysis of such collapsing in the popularization of scientific knowledge for lay audiences in popular science journals).

Next part to talk about and to draw on the large sheet of paper is muscles:

Teacher:	Mm, you have really good muscles. Shall we see if they have a picture of muscles here? I don't know if they have one. Here we are, they do actually have a picture. Can you see, what they have coloured red here, these are meant to be the muscles, which sit outside of the yellow, which is meant to be the skeleton. Do you see? Can you see it too [shows the book to Malin and Richard]?
Richard:	Is it here?
Teacher:	Yes, look, do you see? It's the same there, yes.

Looking at the book, the teacher again makes the children pay attention to the arbitrariness of the colouring of the organs, `Can you see, what they have coloured red here, these are meant to be the muscles, which sit outside of the yellow, which is meant to be the skeleton.` She thematises this issue by making the distinction between what they (the authors, illustrators of the book) `meant`, to what is implied to be how it is. Hence, she hints at the important fact that the representation cannot be taken at face value, simply read off as being an 'adequate' illustration. Some interpretative work is required of the reader.

Having now finished the drawing of the human body, the issue of what, more specifically, this is a representation of, again comes up, as before, in terms of sex:

Teacher:	Look here, shall we write what they are called? Shall we give this fine person a name?
August:	Yes! Yes, he can be called August.
Richard:	No, Richard!
August:	No, August!
Teacher:	Maybe we should find another name, that nobody in this group has?
Richard:	Helen!
Teacher:	That's my name. Can we come up with something that nobody here is called? Malin, do you have any suggestions?
Malin:	Mmmm...
August:	[Inaudible]
Teacher:	Mr, what did you say?
August:	Mr Mästerson.
Teacher:	Mästerson?
Malin:	Kurt.
Teacher:	Kurt Mästerson? Shall we call him that?
Malin:	[Nods]
Teacher:	That's a wonderful name, I think. Is it a man, then?
Children:	Yes.

Asked to name the person on the paper, August and Richard immediately each suggests their own name. Sensing that this will not be solved, the teacher instead suggests that they find a name that is not represented by any of the children in the group. August suggests `Herr Mästerson`, literally Mr. Masterson, to which

Malin adds a first name, Kurt. The teacher and children thus settle the matter with Kurt Mästerson [Kurt Masterson]. Thus, as the teacher asks the children, and they confirm, this is a drawing of a man.

Rounding off the activity, the teacher writes down on the paper, the name of the person depicted as well as the different body parts:

Teacher:	OK! Do you know what I thought I would do first, before we finish? I thought I would write what this actually is, because we may perhaps need to know it and remember it. I'll write it with a fine little pen like this. So let's see. I'll start with this. What was this? Do you remember? [Points to the eyebrows]
Children:	Eyes!
Teacher:	It was eyes, yes. And what was above the eyes.
Malin:	[Points to herself] Eyebrows.

This mode of representing, that is, writing the names of what they have drawn (cf. above) is introduced as a matter of being able to remember at a later time what they have drawn. Hence, per implication, the visual representation is not self-explanatory and something else may be required for the drawing to serve as an external memory (cf. Säljö, 2005) at a later point in time. As a consequence of this practice, the children also get to review and are helped to remember what they have done and spoken about during the event:

Teacher:	August! Mm. What was this? [Points to the heart]
One child:	The heart.
Teacher:	It was the heart.
One child:	It's an A.
Teacher:	It looks almost like an A, yes. What was under the heart here, then? [points to the stomach]
One child:	Uh, a P.
Teacher:	It was where. It was the P, yes. And what was the P, do you remember? It was where the food goes first. Sto...
Malin:	They're my, it's my...
Richard:	Stomach.
Teacher:	Stomach, yes.

When asked to name the objects depicted, the children again use their experiences of the representational system of the alphabet to make sense of what they see. Hence, the heart looks like, or in the child's own terms, is an A. Certainly, a conventionally drawn heart looks like the letter A turned upside down. The teacher also confirms this suggestion as making sense. Asked what is depicted below the heart, one of the children says that it is a P. As introduced by a child earlier when they spoke about the stomach and looked at the illustration in the book, this letter is used in speaking about what something looks like. While the children at first do not seem to remember what the letter represented, that is what it was, instead remembering the symbol in terms of which they perceived it, with some further support by the teacher, Sto..., Richard remembers, stomach.

The activity ends, after half an hour, with the following:

Teacher: Right! I think that Kurt Mästerson is ready, don't you. That's really nice. So we'll talk about the skeleton next time. The thing you talked about, OK [turned towards August].
August: Mm.

The illustration as `really nice (fin)`, that is (also) an object of aesthetic concern, implicating the hybrid nature of this representational practice that we saw earlier is also here touched upon. Finally, the topic of the next event in this theme is mentioned, `So we'll talk about the skeleton next time.` In this way the teacher contextualizes forward, thus supporting the children in being able to connect various events into a narrative, an education in Mercer's (1995, 2008) terms.

10.3.4 Observing and Speaking About a Skeleton

For the third and final activity of the theme, 'The (inside of the) human body', the teacher has a surprise for the children. She has borrowed a full-size skeleton-model of a human being from the local hospital. One of the children immediately says that it is a skeleton.

Malin: It's the skeleton.
Teacher: Yes, exactly. You remember, when you explained about everything that was, that was in the body.
Polly: But, but it's the skeleton. [Creeps forward and points to the skeleton]
Teacher: Yes, that's the skeleton, yes. That's what we have inside here, inside our skin [points to her arm] and our muscles and everything. It's called the skeleton

[---]

Teacher: Let's see, he can sit down here, or she can. Maybe this is Kurt's skeleton? [Teacher says something unclear]. Do you want to come forward?
All the children: Um.
Teacher: Come on! Do you want to feel? This - what do you think this is? [Cannot see what the teacher is pointing at]
Malin: It's the stomach?
Teacher: Yes, can you feel it in here? If you feel here, if you feel on yourselves, you will feel that it is a little bumpy there, right? It is these bones that you feel when you press here on your chest. Inside

	here, what is sitting inside here, do you know?
Eva:	Blood!
Teacher:	There is blood, yes, and something that beats dunk-dunk dunk-dunk.
Malin:	The heart!
Teacher:	The heart is inside there. Do you remember what we breathed with, then?
Polly:	But, but, but Helen, Helen?
Teacher:	Wait, you can ask questions soon. Do you remember what we breathed with? Llll..?
Polly:	Luckiness?
Teacher:	Ll… yes, nearly. Our lungs! That's what we breathe with. Our heart and lungs are inside here.
Eva:	and Kato [unclear] had a really big heart.
Teacher:	Yes, he does, yes.

The teacher's questions and suggestions that the children feel the skeleton and their own bodies relate the model (the physical representation) to its referent (e.g., the ribcage). She also asks a somewhat ambiguous question, `what is sitting inside here, do you know`, and follows up by relating back to the previous occasion and what they talked about (and represented) then, such as the heart and the lungs. Again onomatopoeia works as a scaffold in illustrating and remembering something, `something that beats dunk-dunk dunk-dunk`. This support facilitates one of the children to remember, `the heart!`

In line with her pedagogical work, the teacher is apparently not content with staying at naming the parts of the body. She directs the children's awareness towards functions:

Teacher:	But why do you think we have a skeleton, then?
Richard:	I want to touch the ball.
Teacher:	Do you know? Does anyone have any idea why we have a skeleton? Why do we have it inside our body? Nobody has any suggestions? [The children shake their heads] Shall I tell you something? It's like this, if we didn't have this skeleton – is the skeleton hard or soft?
Polly:	It's hard!
Teacher:	Feel here on the chest. So if we, look here. When he has this hard skeleton, he can stand up [The teacher lifts up the skeleton].
All the children:	Mmm.
Teacher:	Can't he? But if he was completely soft, would he be able to stand up then?
Malin:	No.
Teacher:	With the skeleton we can stand, you see. And we can walk.

The teacher and children talk and think about what difference the skeleton makes to the human body. Hence, the teacher supports the children through her questions, to think about a relevant issue. She is not simply content with naming objects (parts of the body), that is, 'product' issues (what something is) but also process issues (functioning).

The children get to explore the skeleton by touching it and investigating how it looks:

Polly:	The teeth look funny!
Teacher:	Yes, and here we have the teeth. And here I can open up to get to the brain. Let's see! But is there a brain inside there?
Eva:	No.
Teacher:	No, because the brain is not part of the skeleton, but that's where the brain sits, in there. It's inside your heads. The hard bit - if you knock on the head. Whoah! [Polly touches the skeleton's head, causing a piece to almost fall off] Now we have to take care so that it doesn't break, OK? So that the head doesn't fall off.
Polly:	I want to feel inside there.
Teacher:	Of course you can feel.
Polly:	Oooh! What's in there? In here [Feels inside the skull, where the brain should be] what's in here?
Teacher:	It's just to make sure it stays attached there. [A screw inside the skull can be seen in the picture] It has to be secured it with little screws and so on, can you see? Otherwise it wouldn't have stayed in place.

Here an interesting issue concerning representation arises. Polly feeling the inside of the skull, exclaims, `Oooh! What's in there? In here, what's in here?` The teacher explains to her that `It's just to make sure it stays attached there.` Hence, is what the child feels inside the skull a part of the skeleton or merely a part of the model, the representation? As the teacher explains it is the latter. However, this observation is of more general interest. Since any representation means something standing for (representing, illustrating) something else, what is a feature of the representation (e.g., a model) and what is (also) a feature of its referent is always an issue to handle in this kind of learning.

Continuing investigating the skeleton, an interesting simile is introduced and taken up in the talk between the children and teacher:

The children [unclearly]:	The arm.
Teacher:	Yes, can feel inside here that you have this hard bit? It looks just like this. And when we...wait now, we'll see if we can do this.
Polly:	And here are my muscles.
Teacher:	Yes, when you move there, yes, you have muscles, and when we move

	here, the skeleton moves like this.
Richard:	[Takes the skeleton's hand] Good day, good day.
Teacher:	Good day, good day, you say, hello it says back. Do you have any questions about the skeleton here?
Richard:	Yes!
Polly:	These are the muscles. [Points to the ribs]
Teacher:	[To Polly] There are muscles that sit on the skeleton, yes. But what is this, then? [Points to the spine] If you look back here - come on! Come, Richard and Malin, come. If you look back here. What is this, the stomach goes there at the front.
Richard:	It feels like dinosaurs.
Teacher:	Yes, do you know what - I thought the same thing. Our back looks just like that of a dinosaur.
Richard:	And dinosaurs have this!
Teacher:	And feel, if I touch you here [on the back], can you feel that it is a little bumpy there? They're like small bumps, if you feel each other. And those bumps are these [shows on the skeleton].
Richard:	[Unclearly]
Polly:	It's like a person.
Teacher:	Yes, this is a person, yes.
Richard:	Yes, it feels like dinosaur, a dinosaur has these [points to the skeleton]
Teacher:	And you have them too.

Looking at the back of the skeleton and the spinal cord, Richard suggests that `It feels like dinosaurs`. This simile is readily understood by the teacher who confirms that `I thought the same thing. Our back looks just like that of a dinosaur`, and thus at the same time reformulates the simile from how it feels (tactility) to how it looks. Richard excitedly replies, `And dinosaurs have this!` It is commonly known that many children are fascinated by dinosaurs and having seen drawings and perhaps skeletal remains of dinosaurs constitute an important experience for how they make sense of and communicate about what they see, looking at the model of the human skeleton. The teacher returning to the tactile part of exploration, redirects the children's attention to their own and thus the human body, `can you feel that it is a little bumpy there? They're like small bumps, if you feel each other. And those bumps are these` (showing on the skeleton). A new simile is thus introduced in describing what they feel and see, `like small` bumps. Richard maintains his focus on dinosaurs, yes, it feels like dinosaur, a dinosaur has these (showing on the skeleton). The teacher responds, and you have them too. Thus, the teacher works in relating the model to its referent, the human body. The analogy between dinosaurs and the human skeleton provides incentive for this discussion.

Once more the issue of what belongs to the model (the representation) and what belongs to its referent, respectively, comes up in the talk:

Teacher: They are to make sure that everything stays in place. This are the ribs as well, and these are the shoulder blades, and here [shows Malin] are the shoulder blades back here.
Polly: Look here though, what about those ones? Those ones there? [Points under the pelvis]
Teacher: Which ones? Yes, those are also screws so that everything sticks together, otherwise we would have loose bits everywhere. Then we wouldn't have been able to say hello to Kurt here.

10.3.5 Drawing a Skeleton: Negotiating What to Include in the Representation

The teacher now introduces another representational practice to the model of the skeleton and speaking about it. The children are now to draw a skeleton on a paper. The task is introduced in the following way:

Teacher: Look how many fingers Kurt has?
Malin: [Counts the fingers] One, two...
Polly: Why is he called Kurt?
Teacher: Because last week they named the person we are painting over there Kurt, and I thought that this skeleton could also be Kurt. And do you know what we are going to do now? Kurt can sit here, and you are going to...we are going to get some black paper and white pens, because what colour is the skeleton?
All the children: White!
Teacher: You are going to paint your skeleton the way you think it looks inside your body. Just as Kurt has a skeleton, you have a skeleton too.
Malin: Can't we do each other?
Teacher: You're going to paint the skeleton on a piece of paper, but you can paint the skeleton of somebody else if you want. Your Mum's skeleton perhaps?
Many children: Yes!

Teacher: Or your Dad's or Erik's or the baby's skeleton, everyone's skeleton looks like this [points to the skeleton] We also look like Kurt's skeleton.
Child: [Unclearly]
Teacher: I think you should try it.
Richard: Then we could have a yellow one or a blue one.
Teacher: They are actually white, though, skeletons.

The teacher continues speaking about the relation between the skeleton-model and the children, that they themselves look like this inside their bodies. One of the children asks, `can't we do each other?` That is, what is the drawing supposed to be a representation of? For the children this is apparently an important issue, while for the teacher this is of no importance since `everyone's skeleton looks like this. We also look like Kurt's skeleton.` Considering the drawing task, Richard says, `then we could have a yellow one or a blue one.` However, the teacher's response, `they are actually white, though, skeletons`, implies that this time it is not arbitrary which colour to choose and a representation that is in some regards 'correct' is expected of the children (cf. above, the discussions concerning the colours of the illustrations in the book of the inside of the human body).

Teacher: It's like this, today you are going to paint on black paper. Today, I'm choosing. And you are going to paint your white skeletons. That you have.
Eva: But I don't want to paint a skeleton.
Teacher: I think [Unclear, whispering] You can do as much as you feel you can do. You can come and sit here to paint.
Eva: But I don't want to paint a skeleton.
Teacher: But we are trying, just like you can Eva. I think you can all paint beautifully.
Richard: I can!
Teacher: Your skeletons.

Hence, for this drawing task, the children each gets black paper and the white colour. The representational media introduced by the teacher in this case has important affordances and constraints that are well-designed for the present representational purpose. As we have already mentioned, drawing in preschool often takes a hybrid form between 'accurately' depicting something on the one hand and being allowed to draw whatever one wants on the other (see Bendroth Karlsson, 2011, for one such analysis). However, in the present case, as seen in the conversation between one of the children and the teacher, a representation of a skeleton is expected in a certain colour. Still, aesthetics is used as a strategy in encouraging the children to make the drawing. In response to Eva saying, `but I don't want to paint a skeleton`, the teacher responds, `I think you can all paint beautifully.`

The teacher moves the skeleton to the table where the children have taken their seats, so that they can look at it while making their drawings.

```
Polly:     Can I paint the hair?
Teacher:   You can maybe all paint these [points to the ribs].
           He has these ones, you see. And then the chest,
           here. Under the head.
Eva:       But I can do a man, me.
Teacher:   Feel free, paint a man here.
Malin:     I'm painting my Mum!
Teacher:   Are you painting your Mum's skeleton?
Malin:     Mmm.

[---]

Teacher:   Look, Richard has done that too. Look, it looks just
           like that too [points to the skeleton].
Richard:   That's me! That's me!
Teacher:   Is it your skeleton?
Richard:   Yes.
Teacher:   It looks just like that.
```

Being receptive to the teacher's introduction and follow up of the task, Polly asks, `can I paint the hair?` In most drawing tasks this kind of question would not come up, but this drawing activity has been framed in certain representational terms. The teacher's subsequent utterance redirects the children's attention to a part of the skeleton, `you can maybe all paint these` (pointing at the ribs). Clearly the children have understood that not only dinosaurs (see above) but also humans have skeletons, they themselves, their mothers, `a man`.

```
Teacher:   And you can see all these here under the head, then?
           [Points to the ribs] Can you paint them?
Malin:     What are they?
Teacher:   The ribs are these thing here we have in our chest.
           You can feel them here [shows on Malin].
Eva:       You can touch them, do you see [shows on herself].
Teacher:   Yes, you can feel them here.
Polly:     I've painted Kurt's mum [unclear].
Teacher:   Kurt's skeleton there, yes.
Richard:   Look, I've painted there.
Teacher:   Those are the ribs, I can see. Look, wonderful. And
           then you have this long sausage [points and shows
           the spine] which goes in the middle, down to the
           pelvis where the bottom was at the back, wasn't it?
Richard:   Right!
Teacher:   Yes, very good Richard. And then there's this.
Polly:     I've already painted the bottom and the privates!
Teacher:   That's where the privates are, yes.
Polly:     I've painted the privates.
Teacher:   Very good. And then you have the legs here. First
           one long one and then two. And then all the small
           toes. What a lot of bones we have here, don't we.
Polly:     Ready.
```

Malin asks what something she sees on the model is. The teacher names them ribs and shows on the child's own body where they are situated. Eva listens in and says that `you can touch them, do you see`, showing on her own body. The teacher also makes the children attend to another part of the skeleton through pointing at the model and using a metaphor to describe what they look at, `And then you have this long sausage which goes in the middle, down to the pelvis where the bottom was at the back, wasn't it?` Polly exclaims that `I've already painted the bottom and the privates!` [Swedish: `snippan!`] As will Eva later do and say, the girls appear to find it important to represent the sex of the person (the skeleton) they draw (cf. above, when drawing the outlines of a body and whether this is a man or a woman, a boy or a girl). While accepting and supporting the child in this, the teacher continues talking about the bones of the body.

```
Teacher:  Yes, look, how wonderful. And you have done it just
          right, yes Malin. Look. How clever you all are. Eva
          too, really good, look.
Richard:  [Unclearly] Look at me!
Teacher:  Yes, I can see it, exactly. Shall we hold it next
          to it? Here we can see the eyes, and that's the
          mouth with all the teeth, and then you have all of
          these here and then here the long backbone sausage,
          you could call it.
Eva:      And you have to have muscles.
Teacher:  Mm, there are no muscles on the skeleton, do you
          see. We don't actually have them, we can paint them
          another day.
Polly:    Look!
Teacher:  There it is, yes, and there are the arms too. And
          are these the fingers here, do you think?
Polly:    They're the fingers.
Teacher:  They are the fingers, yes, excellent.
```

While being introduced and framed as a representational task, the children's drawings are still valued in aesthetical terms. The metaphor of `the long backbone sausage` returns, now merged with its reference, the spine. One of the girls wants to draw the muscles, but the teacher makes her attend to these not being a part of the skeleton and that they therefore can be drawn another time. Polly's utterance, `look!` is of some interest to our pedagogical concerns. As clarified by developmental researcher, Michael Tomasello (1999), human beings have a unique proclivity to make others attend to what they themselves attend. Two (or more people) sharing attention on something third (in this case a part of the human skeleton) could be considered the very foundation of what we refer to as an education and to pedagogy (cf. Pramling & Pramling Samuelsson, 2010).

The activity and the theme end by the teacher congratulating the children on their `really beautiful skeletons you have drawn now`, and that they should hang their drawings on the wall.

10.4 Discussion: Learning to Represent

Illustrating and analysing these three consecutive learning events on the theme of the human body, with a particular interest in issues of representation, a number of important features have come to the fore. In this final section of the chapter we will summarise these and discuss their importance and implications for children's science learning. Throughout the three episodes we can observe the following features important to children's science learning and representational knowledge:

Children as well as the teacher use metaphors and similes to describe the appearance of what they see, or what the children by the teacher are encouraged to see (`brown peas, like tiny grain of sand, backbone sausage`). This observation is important for several reasons. One thing this use of language implies is the important difference between discerning something and making sense of what is discerned. Learning about nature is not only about noticing certain aspects and phenomena as significant, but also about perspectivising these in relevant and interesting ways. The frequent introduction in this activity (and similar activities as analysed in other chapters of this book) of metaphorical speech and similes also testify to the importance of a teacher and child not only sharing attention (observing the same thing) but also coordinate perspectives on what is observed. The utterances describing what is perceived in terms of similes and metaphors work in establishing a shared perspective. A kind of mutual ground is thus established in the talk between children and teachers. Without any such ground, teachers and children could perceive the phenomenon in entirely different and unrelated ways, making it difficult to contribute to the child's further understanding. The use of metaphors and similes further illustrate the dialectics between everyday concepts and scientific concepts as emphasized by Vygotsky (1987) as necessary for the development of the latter kind of insight.

Another feature of the activity shown by the analysis in this chapter is the importance of learning and distinguishing between representing what one knows is there versus only representing what can be seen from a certain perspective (e.g., `the bottom` in a drawing of a body viewed from the front side). That children often draw what they know is there regardless of whether they can actually observe that feature from where they stand and observe, is well known in developmental literature (e.g., Piaget & Inhelder, 1969; see also Ivarsson & Säljö, 2005, for a discussion and illustration). To draw only what they see, as distinct from what they know is there, is difficult for the children, since it means that they have to disregard what they know. This is no easy task. Still, in terms of developing scientific skills, learning to observe in a closely scrutinized way how phenomena appear under various conditions is an important skill to develop. In science we have theory to not only make sense of what we observe but also to make sense of also what cannot be observed in any straightforward manner (Hanson, 1958/1981). Furthermore, learning to perspectivise phenomena in different ways, and knowing when one or the other perspective is relevant, are important to developing insights into the facts that phenomena can be constituted in language in many different ways and that different traditions of knowledge such as biology, physics, poetry and so forth are premised on certain perspectives. There is no perspective-neutral way of making sense of the world and its phenomena; knowledge is conditioned certain perspectives.

In the analysis of the evolving activities concerning the human body, it was seen how aesthetic preferences reappeared concerning what to represent. This relates to the question of what kind of representational practice this is. Is the purpose to make an accurate representation, accurate for a certain purpose as seen from a certain perspective, or as a kind of art activity? As reported elsewhere, in early childhood settings, drawing is often given the form of a kind of hybrid activity, including both these aims, that is, to represent something 'accurately' and to express oneself 'freely' (see Bendroth Karlsson, 2011; Kress, 1997; Pramling & Pramling Samuelsson, 2011)? What kind of activity the making of drawings is, is seen in the teacher's introduction and responses to the children's suggestions and drawings during the activity. Making the drawings of the human body appear to be a multi-purpose activity. This is also a way of coordinating the children's sense making and interests and the intention of the teacher in making children aware of and understanding certain things.

Closely related to the issue of what kind of activity the drawing session that comes to the fore in these activities is what more specifically a representation should be a representation *of* (e.g., a human being, a man, a woman, a certain person such as the child him- or herself making the drawing, his or her mother, father, etc.). While for the teacher it does not matter which human being is illustrated, since the point is to learn about 'the human body', to the children this clearly makes a difference. The children relate the task to persons that are familiar and important to them, for instance a family member. To some extent, the issue of which human body is depicted is relevant to the task and it comes up for negotiation in terms of whether the body is a female or a male body. Some children also stay at this difference and integrate it into their drawing of the skeleton, resulting in a drawing that is a kind of hybrid.

Looking at the teacher in the followed activities, we can see how she makes many important things. For example, when a child introduces a certain metaphor, she confirms that what they talk about looks like this (i.e., simultaneously implying that it is not, in fact this) and says what it is. In supporting and clarifying the child perhaps thinking that it looks like this, the teacher motivates the child's suggestion as a relevant contribution but also adds something to further his or her knowledge. Metaphors and similes are representations in speech. An important distinction to clarify is therefore what something is and what is looks like, that is, how the utterance relates to its referent. Important as it is to learn to make such a distinction, when learning about phenomena, using the former, that is, what something looks like, to make sense of the latter, that is, what it is, shows the dialectics between everyday and scientific concepts (Vygotsky, 1987), as we elaborated on in Chap. 9. Learning that what something looks like is distinct from what it is, is important, but the former can be used as a resource in learning the latter, if it is thematised in conversation with for example a teacher. As seen when the second event commences, the teacher departs from a child's recollection of the previous occasion and expands it further, as in this case typically going from what something is called to what function it fills for us, our human body. The teacher also makes important connections between representations (in the book and on their own drawings) to their own bodies, clarifying how the representations relate to their referents (e.g., veins). This is an important recontextualisation, weaving together the novel with children's experiences and also how different representations can be representations of the same thing, but in different ways.

The teacher also thematises what is a part of the representation as such (e.g., the brain being depicted by a blue colour in the book) as distinct from the nature of its referent. However, this colouring, which may be reasonable for illustrative purposes (making it easier to discern different organs) and aesthetic purposes, does seem to pose some difficulties to some children. As seen in the excerpt, one of the children responds to the teacher's question about what colour she thinks the brain is by saying blue, while other children give other suggestions (green, red). However, something to bear in mind is that the teacher's initial question is not unambiguous, and could in fact be taken as asking about the brain depicted in the book (which is actually blue). Hence, it is not clear that the children mistake the representation from its referent. Still, there is a potential problem even for older children in science education learning how to take representations such as models and other graphical depictions. For example, as reported by Molander, Pedersen and Norell (2001, p. 206), in their study from compulsory school, a student may reason about an atom in the following way: "there is something in physics, something to do with atoms [...]. Something red and white and black.. some sort of ball". In addition to seeing how the student makes sense of the concept in more familiar terms (colours, ball), this reasoning points at the problem in learning to distinguish between a representation (how the phenomenon is mediated) and what it represents. The learner is faced with the issue of what features of the representation to consider relevant, for example, at different levels of description (e.g., atomic level and the level of the representation, respectively) (Pramling, 2006a). Another example of this issue is when a child asks what something is inside the skeleton and the teacher clarifies that it is only screws to make sure it holds together, and therefore not a part of the skeleton as such.

The teacher also, through her responses to the children's metaphors and similes, challenges the children to clarify what sense they make. For example, on one occasion a child suggests that the intestines looks like a `tooth brush`, which does not make sense to the teacher, as evident in her response. When not being able to establish temporarily sufficient intersubjectivity (Rommetveit, 1974; see also Chapter 11) with the teacher, allowing them to go on with the activity, the child responds by reformulating that the intestines look like `tooth paste`, which does make sense to the teacher. Through adjusting one's communication in this manner to an interlocutor, the child is socialized into attending to what he or she needs to make explicit to make sense to someone else and what can be left implied. This, of course, also implies that others may not understand and see the world as I do (see also in Chapter 11, where we discuss this matter).

Another interesting observation from the studied activities that cuts to the heart of the theme of the present book, is how children's perception is evidently semiotically mediated (Wertsch, 2007) by their cultural knowledge. An example of this is when the children report seeing intestines as looking like certain letters of the alphabet (P, A). Growing up in a literate culture with this system also shapes the children's perception (Olson, 1994). In his fascinating study, and one of the classics of cultural-historical research, Alexander Luria (1976) report findings that illustrate how learners' perception (as well as other important cognitive and communicative functions such as categorizing and reasoning) change when they become participants in novel activities, in this case, in novel institutional arrangements. He studied what happened to adults who were allowed to attend school that had recently opened in an area previously without such an institution. Without going into the details of his extensive and rich study, for the present discussion, his work is important among other things for showing how even how we perceive the world and its phenomena changes, that is, is learned and that how we learn to perceive depends on what practices and cultural tools we are introduced to and come to appropriate (cf. Kozulin, 1998; Wertsch, 1998). These cultural tools will come to semiotically mediate (Wertsch, 2007) phenomena for the learner. Learning to see in institutionally relevant and expected ways means to perceive in terms of particular tools (Goodwin, 1994). As we have already pointed out above, when discussing the difference between a child drawing what he or she knows and what he or she can actually see from a certain perspective, the basis of scientific observation, that is, seeing should not be taken for granted as unproblematic to science education. Seeing in this context entails more than meets the eye.

The teacher is further important in recontextualising (van Oers, 1998) backwards and forwards (in addition to how they do so between representations and children's experiences, as we have already mentioned) between events. Through this 'weaving' (cf. the etymology of 'text' as writing and weave, Barnhart, 2000) what would otherwise risk becoming separate events or phenomena for the children are turned into what Mercer (1995, 2008) refers to as an education. An education, according to this notion, is more than simply a number of things learned (fragmentary facts). Rather, it presumes and consists of some kind of connected construal, a narrative of some sort that makes these meaningful in relation to, and in light of, one another. Such a relation is necessary to create continuity and thus cumulativeness in learning beyond simply learning different things. Notably, the children also recontextualise what they look at and speak about. In this case, they made sense in terms of referring to fictional stories (such as stories by Astrid Lindgren).

References

Änggård, E. (2005). *Bildskapande: En del av förskolebarns kamratkulturer* [Making pictures: A part of preschool children's peer cultures] (Linköping studies in arts and science, 315). Linköping, Sweden: Linköping University.

Aronsson, K., & Hundeide, K. (2002). Relational rationality and children's interview responses. *Human Development, 45*, 174–186.

Barnhart, R. K. (Ed.). (2000). *Chambers dictionary of etymology*. Edinburgh, Scotland: Chambers Harrap.

Bendroth Karlsson, M. (1996). *Bildprojekt i förskola och skola: Estetisk verksamhet och pedagogiska dilemman* [Arts projects in preschool and school: Aesthetic activity and educational dilemmas] (Linköping studies in arts and science, 150). Linköping, Sweden: Linköping University.

Bendroth Karlsson, M. (2011). Pictures of spring: Aesthetic learning and pedagogical dilemmas in visual arts. In N. Pramling & I. Pramling Samuelsson (Eds.), *Educational encounters: Nordic studies in early childhood didactics* (pp. 85–104). Dordrecht, The Netherlands: Springer.

Carey, S. (1985). *Conceptual change in childhood*. Cambridge, MA: MIT Press.

Cuthbert, A. J. (2000). Do children have a holistic view of their internal body maps? *School Science Review, 82*, 25–32.

Daniels, H. (2005). Vygotsky and educational psychology: Some preliminary remarks. *Educational & Child Psychology, 22*(1), 6–17.

Donaldson, M. (1978). *Children's minds*. Glasgow, Scotland: Collins.

Fleer, M. (2009). Supporting scientific conceptual consciousness or learning in a 'roundabout way' in play-based contexts. *International Journal of Science Education, 31*(8), 1069–1089.

Goatly, A. (1997). *The language of metaphors*. London: Routledge.

Goffman, E. (1974). *Frame analysis: An essay on the organization of experience*. New York: Harper & Row.

Goodwin, C. (1994). Professional vision. *American Anthropologist, 96*(3), 606–633.

Guichard, J. (1995). Designing tools to develop the conception of learners. *International Journal of Science Education, 17*(2), 243–253.

Hanson, N. R. (1981). *Patterns of discovery: An inquiry into the conceptual foundations of science*. New York: Cambridge University Press. (Original work published 1958)

Hayes, D., Symington, D., & Martin, M. (1994). Drawing during science activity in the primary school. *International Journal of Science Education, 16*(3), 265–277.

Hundeide, K. (1977). *Piaget i kritisk lys* [Piaget in a critical light]. Oslo, Norway: Cappelen.

Ivarsson, J., & Säljö, R. (2005). Seeing through the screen: Human reasoning and the development of representational technologies. In P. Gärdenfors & P. Johansson (Eds.), *Cognition, education and communication technology* (pp. 203–222). Hillsdale, NJ: Lawrence Erlbaum.

Kozulin, A. (1998). *Psychological tools: A sociocultural approach to education*. Cambridge, MA: Harvard University Press.

Kress, G. (1997). *Before writing: Rethinking the paths to literacy*. London: Routledge.

Kress, G. (2003). *Literacy in the new media age*. London: Routledge.

Luria, A. R. (1976). *Cognitive development: Its cultural and social foundations* (M. Lopez-Morillas & L. Solotaroff, Trans.). Cambridge, MA: Harvard University Press.

Mercer, N. (1995). *The guided construction of knowledge: Talk amongst teachers and learners*. Clevedon, UK: Multilingual Matters.

Mercer, N. (2008). The seeds of time: Why classroom dialogue needs a temporal analysis. *Journal of the Learning Sciences, 17*(1), 33–59.

Molander, B.-O., Pedersen, S., & Norell, K. (2001). Deaf pupils' reasoning about scientific phenomena: School science as a framework for understanding or as fragments of factual knowledge. *Journal of Deaf Studies and Deaf Education, 6*(3), 200–211.

Olson, D. R. (1994). *The world on paper: The conceptual and cognitive implications of writing and reading*. Cambridge, UK: Cambridge University Press.

Osbourne, J., Wadsworth, P., & Black, P. (1992). *Processes of life: Primary SPACE project research report*. Liverpool, UK: Liverpool University Press.

Óskarsdóttir, G. (2006). *The development of children's ideas about the body: How these ideas change in a teaching environment* (Diss.). Reykjavík, Iceland: University of Iceland, Faculty of Social Sciences.

Piaget, J., & Inhelder, B. (1969). *The psychology of the child*. London: Routledge & Kegan Paul.

Pramling, N. (2006a). *Minding metaphors: Using figurative language in learning to represent* (Göteborg studies in educational sciences, 238). Göteborg, Sweden: Acta Universitatis Gothoburgensis.

Pramling, N. (2006b). 'The clouds are alive because they fly in the air as if they were birds': A re-analysis of what children say and mean in clinical interviews in the work of Jean Piaget. *European Journal of Psychology of Education, 21*(4), 453–466.
Pramling, N., & Pramling Samuelsson, I. (2010). Evolving activities and semiotic mediation in teacher-child interaction around simple objects. *Educational & Child Psychology, 27*(4), 22–30.
Pramling, N., & Pramling Samuelsson, I. (2011). En korpus av kunskap – förhandling mellan lärare och barn om att representera den mänskliga kroppen i en bok. In V. Glaser, K. H. Moen, S. Mørreaunet, & F. Søbstad (Eds.), *Barnehagens grunnsteiner. Formålet med barnehagen* (pp. 176–187). Oslo: Universitetsforlaget.
Pramling, N., & Säljö, R. (2007). Scientific knowledge, popularization, and the use of metaphors: Modern genetics in popular science magazines. *Scandinavian Journal of Educational Research, 51*(3), 275–295.
Reiss, M. J., Tunnicliffe, S. D., Andersen, A. M., Bartoszeck, A., Carvalho, G. S., Chen, S. Y., et al. (2002). An international study of young peoples' drawings of what is inside themselves. *Journal of Biological Education, 36*(2), 58–64.
Rommetveit, R. (1974). *On message structure: A framework for the study of language and communication*. London: Wiley.
Säljö, R. (2005). *Lärande och kulturella redskap. Om lärprocesser och det kollektiva minnet* [Learning and cultural tools: On processes of learning and collective memory]. Stockholm: Norstedts.
Säljö, R., & Wyndhamn, J. (1993). Solving everyday problems in the formal setting: An empirical study of the school as a context for thought. In S. Chaiklin & J. Lave (Eds.), *Understanding practice: Perspectives on activity and context* (pp. 327–342). New York: Cambridge University Press.
Thulin, S., & Pramling, N. (2009). Anthropomorphically speaking: On communication between teachers and children in early childhood biology education. *International Journal of Early Years Education, 17*(2), 137–150.
Tomasello, M. (1999). *The cultural origins of human cognition*. Cambridge, MA: Harvard University Press.
Tunnicliffe, S. D., & Reiss, M. J. (1999). Students' understandings about animal skeletons. *International Journal of Science Education, 21*(11), 1187–1200.
van Oers, B. (1998). From context to contextualizing. *Learning and Instruction, 8*(6), 473–488.
Vygotsky, L. S. (1978). *Mind in society: The development of higher psychological processes*. Cambridge, MA: Harvard University Press.
Vygotsky, L. S. (1987). *The collected works of L. S. Vygotsky, Volume 1: Problems of general psychology, including the volume* Thinking and Speech (R. W. Rieber & A. S. Carton, Eds., N. Minick, Trans.). New York: Plenum.
Vygotsky, L. S. (1991). *The psychology of art*. Cambridge, MA: MIT Press.
Vygotsky, L. S. (1997). *The collected works of L. S. Vygotsky, Volume 4: The history of the development of higher mental functions* (R. W. Rieber, Ed., M. J. Hall, Trans.). New York: Plenum Press.
Vygotsky, L. S. (1998). *The collected works of L. S. Vygotsky, Volume 5: Child psychology* (R. W. Rieber, Ed., M. J. Hall, Trans.). New York: Plenum.
Wells, G. (2007). Semiotic mediation, dialogue and the construction of knowledge. *Human Development, 50*(5), 244–274.
Wertsch, J. V. (1998). *Mind as action*. New York: Oxford University Press.
Wertsch, J. V. (2007). Mediation. In H. Daniels, M. Cole, & J. V. Wertsch (Eds.), *The Cambridge companion to Vygotsky* (pp. 178–192). New York: Cambridge University Press.

Chapter 11
The Nature of Scientific Educational Encounters

Niklas Pramling

Abstract In this chapter the nature of the educational encounter is presented. The educational encounter is conceptualized primarily as an interaction between an adult (e.g., a preschool teacher, acting as a more experienced peer) and a child. How the encounter plays out in a concrete sense is critical for understanding what opportunities children are given and what kind of support they receive during their learning of science. A cultural-historical perspective on learning and development shows that communicative support is fundamental to a child's development and this perspective is different from an exploratory-based notion of children's development where they develop by their own accord as they explore the world. While an exploratory view could explain some learning, it is insufficient to explain more abstract forms of knowledge such as typical scientific knowledge. The chapter also discusses how not all encounters between two (or more) people can be viewed as an educational encounter. The idea of a scientific encounter with distinctive features is introduced in this chapter.

Keywords Interactions • Mediation • Discovery learning • Sustained shared thinking • Deictic referencing • Intersubjectivity

11.1 Introduction

In this section of the book we will summarise and discuss several important features of the educational activities that we have analysed in previous chapters. The overarching theme for this chapter is the nature of the educational encounter, primarily between an adult (e.g., a preschool teacher, acting as a more experienced peer) and a child. We will argue that how this encounter plays out in a concrete sense is critical to what children are given the opportunity and support in developing. Taking a cultural-historical perspective on learning and development, communicative support is considered fundamental to a child's development (Mercer & Littleton, 2007). This is a rather different perspective than an exploratory-based notion of children's development, that is, the idea that children develop by their own accord as they explore the world. While an exploratory view

M. Fleer, N. Pramling, *A Cultural-Historical Study of Children Learning Science*, Cultural Studies of Science Education 11, DOI 10.1007/978-94-017-9370-4_11

could explain some learning, it is insufficient to explain more abstract forms of knowledge such as typical scientific knowledge. In addition, it should be clear what we refer to as an 'educational encounter'. Following Pramling Samuelsson and Pramling's (2011) definition, not any encounter between two (or more) people is an educational one, but a scientific encounter has certain distinctive features. What features these are will shortly be explained and illustrated.

The features we will point out and discuss in this chapter are the following: the distinctive features of scientific 'educational encounters' (Pramling Samuelsson & Pramling, 2011), including 'sustained shared thinking' (Siraj-Blatchford, 2007); deictic referencing and the linguistic informing of experience; how to avoid the pitfalls of 'illusory intersubjectivity' (Ivarsson, 2003); the difference between exploratory (Piaget, 1970) and teacher mediated learning (Wells, 1999); how the variety in understanding among a group of children can be used as an asset and pedagogical principle in developing children's knowledge (Pramling, 1994, 1996); and the distinction and relational management of everyday and scientific concepts (Vygotsky, 1987).

We specifically introduce these general pedagogical concepts in order to discuss in the latter part of the chapter, discovery learning in science.

11.2 Educational Encounters

In a recent volume on children's learning in early childhood education settings (primarily Swedish and Norwegian preschools), Pramling Samuelsson and Pramling (2011) summarise the features of what they refer to as an 'educational encounter'. It is decisive to realize that what is here referred to as an 'education' is not the same as 'learning'. The latter concept is far more general, and obviously children and others learn a great deal without being enrolled or engaged in any activity that would be referred to as an education. Hence, when Pramling Samuelsson and Pramling write about educational encounters, they have certain institutional arrangements in mind. The defining features of such arrangements are an interest in and an ambition to build upon the children's perspectives, trying to establish and maintaining temporarily sufficient intersubjectivity (Rommetveit, 1974), through recontextualising and meta-communication establish an education from a series of events, teachers introducing and scaffolding children to appropriate 'the tools of the domain' (e.g., distinctions and categories), and coordinating the children's perspectives and the perspective of the domain. In connection to these educational features, we will now discuss early childhood science education.

11.2.1 Establishing Intersubjectivity or Sustained Shared Thinking

A popular concept coming out of the large-scale EPPE project in the UK (Sylva, Melhuish, Sammons, Siraj-Blatchford, & Taggart, 2010), is 'sustained shared thinking'. Siraj-Blatchford (2007) explains this notion in the following terms:

> The EPPE Qualitative analysis revealed a general pattern of high cognitive outcomes associated with sustained adult-child verbal interaction along with a paucity of such interactions in those ECE settings achieving less. 'Sustained shared thinking' (Siraj-Blatchford et al., 2003) thus came to be defined as: '...an effective pedagogic interaction, where two or more individuals "work together" in an intellectual way to solve a problem, clarify a concept, evaluate activities, or extend a narrative'. (p. 17f.)

In relation to this concept, Siraj-Blatchford (2007) also makes a point about pedagogy and early childhood education that is highly relevant to our present discussion. She writes that: "As I have argued elsewhere (Siraj-Blatchford, 1999) any adequate definition of pedagogy for early childhood education must include the indirect scaffolding provided by adults in e.g., providing the stimulating learning environments for socio-dramatic play." While what she points out as a distinguishing mark of such a practice may be necessary, is it also sufficient? An 'education' as opposed to the more general concept of 'learning' in Pramling Samuelsson and Pramling's (2011) understanding also includes a more competent communicative partner who introduces and scaffolds the appropriation of some important cultural tools (e.g., categories, distinctions) and who also supports the child in recontextualising (van Oers, 1998) activities into a coherent whole; the latter is according to Mercer (2008) what constitutes an education from a number of events. Hence, it is worth keeping in mind when we speak about 'early childhood education' and when we speak of the more general notion of 'children's learning'. Since the discussion of the present chapter is on educational models (early childhood education), this is important to consider.

If we return to the concept of 'sustained shared thinking', a perhaps more familiar term for this phenomenon is 'intersubjectivity'. However, the latter term has been understood in many different ways in different traditions of thinking. Importantly, scholars building on the work of the later Wittgenstein have emphasized that intersubjectivity does not mean that two or more interlocutors have identical concepts. Rather, intersubjectivity is a temporarily shared focus of attention making it possible for interlocutors to go on with a shared activity (Rommetveit, 1974), as distinct from pursuing diverse and parallel one another, lines of inquiry. An illustration of the latter can be found in a study by Ivarsson (2003) on computer-assisted learning. Investigating the notion of 'recursion', children and their teacher were able to interact around a computer program using deictic references such as pointing and using words such as 'there', 'that', etc. However, while these references signified conceptual distinctions for the teacher, there was no indication in the children coming to such an understanding. Rather, they manipulated buttons without a 'deeper' conceptual understanding. Ivarsson (ibid.) labels the activity as illustrating 'illusory intersubjectivity', that is, children and teacher in one sense refer to the same objects but conceptually these are distinct matters for the communicative partners. Another illustration of the difficulty of establishing 'temporarily sufficient intersubjectivity' (Rommetveit, 1974) can be found in Säljö, Riesbeck, and Wyndhamn's (2001) study of group work on elementary geometry (the triangle as a geometric object and how to calculate its area) in Swedish primary school. One of the points made by their analysis is that the children and their teachers were not coordinated in their communication. Significantly, the children used the Swedish word 'trekant' (literally: 'three-angle') while their teachers used the geometrical term 'triangel' ('triangle'). This may seem like synonymous terms. However, these illustrate the important difference between what Vygotsky (1987) referred to as 'everyday concepts' and 'scientific concepts'. While 'three-angle' is functional for the children in solving the task of cutting out this shape from a paper, it does not relate systematically to other concepts, like the geometrical concept 'triangle' does, allowing them to calculate its area. Hence, the lesson goes on, but the intersubjectivity is, in Ivarsson's terms, 'illusory'. Säljö and colleagues reason that what they have analysed in their study (group work) is a common form of organising learning in classrooms. They further suggest that this form of education which is sometimes referred to as 'pupil active' or 'pupil steered' is very much a heritage from Piagetian theory. According to this perspective, the child's understanding will be a result of his or her independent manipulation and observation of the world (ibid.). However, as they conclude on the basis of their empirical study:

> From a Piagetian perspective, we could say that an intended accommodation does not appear. The pupil does not change his/her mental structure so that new information can be attached. The pupil does not understand the world in a new way. To see and to do are no guarantee for understanding. [...] The teachers are notably insensitive to this fact and only reluctantly take part in the pupils' conversations. In the passages we have registered, the teachers have difficulties to achieve and sustain a mutual perspective with the pupils on problems. (p. 236, our translation)

In contrast to such a perspective on children's learning, Säljö et al. (2001) clarify how their findings can be interpreted from a sociocultural (cultural-historical) perspective:

> From Vygotskian points of view, we could instead say that the pupil appropriates new knowledge first through reworking and working through different interpretations of the practical work. Cooperation in the form of a 'negotiation' with the teacher or another peer [...] paves the way for new insights. This requires coordination or in other words a shared perspective and an adequate language with which to speak about what the physical material shall illustrate. A clear discourse must be established. (p. 236f., our translation)

Säljö et al. (2001) draw a number of conclusions. First, that so-called pupil-active or pupil-lead activities, while in some sense may be necessary, are not sufficient in order to develop the children's understanding. Second, in order to make use of the practical work and concrete observations the lessons revolve around, at least two additional features are necessary: (a) the coordination of perspectives (between the teacher and the pupils), so that they can agree in what way and in what terms to speak about the object of inquiry and (b) the teacher introducing and scaffolding the pupils in using a certain language (a discourse, in this particular case, a geometrical discourse). Even when the activity is guided by practical manipulation of concrete objects, the participation of the teacher far beyond providing sufficient material is necessary in order to support children developing the more abstract forms of knowledge Vygotsky (1987) refers to as 'scientific concepts'. Obviously this last point is inherently intertwined with curricula. If the intent is for the children to 'get a feel for', in this case, geometrical shapes, then the conclusions here drawn from a Vygotskian perspective would not be relevant. However, in the present case, the Swedish curriculum prescribes an intention with the children's knowledge development of a 'scientific' (in the Vygotskian sense) kind. The role, if any (cf. Siraj-Blatchford, 2007), of a teacher in early childhood education is contingent upon the framing provided and promoted by guiding documents such as a curriculum.

11.2.2 Mediated Learning

While the concept of 'scaffolding' was not used by Vygotsky himself, it has been a frequently employed concept within cultural-historical theory since it was introduced in a seminal paper by Wood, Bruner, and Ross in 1976. In their study, Wood et al. (1976) analysed adult-child conversation and interaction when engaged in carrying out a problem-solving task. Through their analysis, they were able to show how adult and child changed the division of labour in solving the task, that is, who did what and how this changed during the course of the activity. That a 'more experienced peer' (Rogoff, 1990, 2003) provide some support, structuring resources (Lave & Wenger, 1991), or in Wood et al.'s terms, scaffolding, and that this support changes as the child come to take over increasingly more responsibility for the different steps of the problem-solving activity are important to understand children's learning in interaction and communication with others.

In the course of theorizing, the concept of scaffolding has been critically scrutinized. In a review of this critique, Stone (1998) summarizes the most important critique as revolving around the following issues: that this model of interaction may be culturally specific, that it emphasizes the micro-level of analysis rather than macro-level issues of child development, focuses adult-child interaction rather than child-child interaction, that this kind of interaction may not be frequent in children's lives, and that discussions about scaffolding has been less specific about the mechanisms. However, it could be argued that the focus the concept of scaffolding places on the micro-level of analysis is necessary for understanding how children are assisted in learning, and that how this interaction plays out in a concrete sense is a legitimate interest for research on learning and development. Whether or not focus is on adult-child interaction rather than child-child interaction is a matter of what kind of situations are studied, rather than a feature of the concept as such. In fact, as Stone also points out, there is also research on child-child interaction in this vein. In principle, any more experienced peer could scaffold another child's development. As for the argument that scaffolding may not be frequent in children's lives, this cannot be seen as a critique of the usefulness and value of the concept for studying certain educational activities. Finally, the argument that researchers have not always been specific about the mechanisms of scaffolding, this may be the case but it is not true of the original conception as reported in Wood et al.'s (1976) study, where they do clarify in detail what this assistance consists of in the activity they follow. In fact, clarifying what scaffolding means in a more concrete sense in various activities is of considerable interest to research on children's learning and development. Like any concept in science, the value of scaffolding needs to be decided on the basis of what one intends to say something about. If wanting to investigate more specifically what the changing division of labour between, for example, a teacher and a child consist, then this concept may indeed be useful, as it has been in many studies.

However, over time, the term 'scaffolding' has spread to discussions about education in, for example, policy documents, which are not theoretically grounded and elaborated. This has perhaps made the concept somewhat vague. However, the same argument could be made concerning other theoretical terms such as 'mediation' that is often used simply as 'teacher mediation' rather than the more theoretically crucial notion of 'semiotic mediation' (Vygotsky, 1987; Wells, 2007; Wertsch, 2007) and has been positioned in the general literature as being integral to understanding a Vygotskian perspective. And, of course, the very term 'learning' which is conceptualized in a particular way within cultural-historical theory – as the appropriation of cultural tools and practices (Rogoff, 1995; Tomasello, 1999; Wertsch, 1998) – is used in many different ways for various purposes. "Given its attractiveness," Mercer (1995) writes,

> it is not surprising that the term 'scaffolding' is now commonly used in educational research and by teachers discussing their own practice. However, I have some reservations about its being casually incorporated into the professional jargon of education, and applied loosely to various kinds of support teachers provide. The essence of the concept of scaffolding as used by Bruner is the sensitive, supportive intervention of a teacher in the progress of a learner who is actively involved in some specific task, but who is not quite able to manage the task alone. Any other kinds of help provided by teachers are better described as 'help'. (p. 74)

Mercer further writes that the reasons for him questioning the usefulness of 'scaffolding' for conceptualizing school practices are, for example, teacher-child ratios as fundamentally different from the dyadic relationships originally referred to by the concept. He argues that "A theory of the guided construction of knowledge in schools cannot be built upon comparisons with teaching and learning in other settings. To be useful, the concept of 'scaffolding' must be reinterpreted to fit the classroom" (ibid., p. 74). "Education", he argues, "is not about the physical manipulation of objects" (p. 74). Rather, "A great deal of it is learning how to use language – to represent ideas, to interpret experiences, to formulate problems and to solve them" (p. 74f.). Connecting to this discussion, in a later account, Wells (1999) suggests, that "one of the chief functions of the use of language in the classroom is to induct students into modes of discourse that provide them with frames of reference with which to 'recontextualize' their experience, and that it is this task that gives educational scaffolding its particular character" (p. 127; cf. Mercer, 2000).

In a study similar to Nilholm and Säljö's (1996) study of Swedish mother-child dyadic problem solving (cf. also Wertsch, 1979), Sun and Rao (2012) compared the scaffolding of Chinese mothers and teachers, respectively, in dyadic problem-solving activities with kindergarten children (approximately 5 years old). In their study, Nilholm and Säjö studied problem-solving dyads with mothers and their 6-year-old child. The problem was to tie a knot (a clove hitch) using a schematic picture as a resource. The mothers differed in terms of education and profession (industrial workers, nurses and teachers). Briefly, Nilholm and Säljö found many similarities between the groups but one difference was that "the teacher mothers were more inclined to involve the child as a performer and to organize the cooperation in such a way that the child had to engage in the semiotic activity of relating the picture to the tying of the rope" (p. 325). The researchers explain this difference in terms of the participant mothers' definition of the task and what it means to learn in such a situation. Sun and Rao studied how an adult and child solved four different tasks: supermarket (buying a combination of fruit with a certain amount of money, do a jigsaw puzzle, an arithmetic task, and a map problem). The interactions between the adult and the child were videotaped and analysed in terms of how the activity developed. It was found that "teachers gave higher-level cognitive support and emotional feedback than did mothers" (p. 246). The mothers differed in that those "with more education provided more optimal scaffolding than those with less education" (ibid.). The teachers did not tend to adjust their scaffolding to the two groups of children, that is, those children with more respectively less educated mothers. Both teachers as well as the mothers adjusted how they scaffolded the child's problem solving in response to the characteristics of the task. One important finding of the study was that "professional training in early childhood education is important for equipping adults with effective scaffolding skills" (p. 260). More specifically, "teachers showed a higher level of scaffolding manners, less negative feedback, and transferred more responsibility to children than mothers" (ibid.). Another important finding of these studies is that they show in a rather concrete sense how children are given different developmental opportunities due to the varying participation of adults in joint activities. Scaffold a child to solve a problem does not merely refer to making sure the problem is solved in the present situation. Rather, the concept entails that the child will successfully take more active part in carrying out this form of problem solving, and similar ones, in subsequent situations. Hence, the premise is that through participating in activities where another regulates one's activities, the child will come to develop self-regulative capacities (see Wertsch, 1979, for an elaboration on this Vygotskian idea).

As we have already mentioned, the concept of scaffolding has received some critique, for example, by scholars such as Mercer (1995) arguing that the concept originally referred to a situation of one-to-one interaction (Wood et al., 1976) and that it therefore is perhaps not useful for understanding learning in classrooms where one teacher rarely interacts with one child at a time for a sustained time. While this is certainly true, the basic idea of the metaphor of scaffolding as changing division of labour between interlocutors points at an important feature of learning in many situations, including learning in educational settings. The concept of scaffolding as used in this theoretical tradition does not simply mean 'support' of any kind, but a gradual change in division of labour between participants. It thus, among other things, serves to highlight the important contributions made by others, such as a teacher, to the child's learning, which is important to understand learning in educational institutions such as preschool and school. It is important to remember that 'scaffolding' is a metaphor. Like all metaphors it mediates our perception and cognition, that is, it 'informs' and directs our attention. It is useful since it provides a means of conceptualizing the important role of a more experienced peer, such as an early years teacher, in the child's development. Since phenomena such as 'learning' and 'understanding' are not directly available to inspection, we need metaphors to talk about these. However, it is an important theoretical discussion to keep alive, what metaphors to use when studying and conceptualizing children's learning and development. We also discuss this point in Chap. 9.

11.2.3 Using Children's Different Understanding as a Resource and Pedagogical Principle

In any group of children there will be a variation in ways of understanding a phenomenon or a theme that is being investigated and talked about. In a series of studies, Ingrid Pramling (1990, 1994, 1996) has shown how this basic empirical fact can be used as a pedagogical principle in developing children's understanding. One example is the making of children's song sheets in order to remember which songs to sing at an upcoming cultural event, the celebration of Lucia (13th December each year). Lucia is the bringer of light in a dark time of the year and she is celebrated through a so-called Lucia-procession where children with electric or live candles in their hair walk into a dark room singing traditional songs for the occasion, usually before the invited parents. This is a common cultural practice in Swedish preschools and schools. While children making song sheets for this event may seem an odd example within the framework of the present book, what concern us here are mainly two things. First that the children are given the task of representing an event on paper (an issue we study in detail in Chapter 10). Second, this way of working, as we will now describe, illustrates how the variety among a group of children's understanding can be used as an educational principle and asset in furthering their development.

The reason for the teacher encouraging the children to represent the song repertoire on paper is that the children can have difficulties remembering what songs to sing and in which order. As described by Pramling Samuelsson and Asplund Carlsson (2003), the teacher first gives the children the task of dividing their paper (through drawing) into twelve frames. The reason for this number is simply that the children will be singing 12 songs on the upcoming occasion. As a consequence of this task, the children get to solve a mathematical problem. However, the teacher's main objective is to allow the children to reflect on writing (graphical representation). The children and teacher then sing the first song together. Having done so, the children are encouraged to write and/or draw a symbol for the first song in the first frame. The teacher asks them to think about what the song is about and how it can be drawn or written in a way that they can remember what song it is. In order to remember in what order to sing the songs, the teacher also suggests the children to think about how they can know in what order the songs come. In response to this question, the children variously use numbers and letters. This sequence is then repeated with the other songs they are going to sing. Even if the resulting song sheets are unique for each child, this difference is made explicit and discussed among the children and teacher, that is, the children's solution to the problems of remembering what songs to sing and in what order, are made into a topic of discussion. The purpose of this activity is to make the children aware of the fact that problems can be solved in different ways and not everyone does the same. To learn that not everyone understands the same way as oneself does is a very important lesson in life. In addition, to discuss to what extent a representation is intelligible also to another child could be the next step in their development, thus introducing the insight that in order to serve as an external memory (Middleton & Edwards, 1990; Säljö, 2005) also for someone else, either some kind of depiction or conventional sign would perhaps be necessary. In Chapter 10 of this book, we could observe how the relationship between idiosyncratic representations and more conventional ones came up for negotiation in the talk between children and their teacher.

11.2.4 Discovering by Oneself or Mediated Through Communication

If we return to the discussion referred to above to Säljö et al. (2001) study between different concepts of learning, what they referred to as a Piagetian notion based on exploration and discovery and a Vygotskian notion based on mediated activity, we can further emphasise and illustrate this important difference in how to account for children's development. The Piagetian notion of development has been very influential for how educational experiences are organized in many parts of the world. In a description of the manifestation of this view, Säljö writes (on school, but basically the same argument could have been made about early childhood education settings such as preschool):

> When entering a classroom today in many European countries, but also in many other places around the world, the chances are great that you will enter an environment that is heavily inspired by Piagetian notions of teaching (see e.g., Bergqvist, 1990, for an insight into Swedish teaching and Edwards & Mercer, 1987, for British conditions). Curricula and similar official documents formulated in the 60s, 70s, 80s and 90s in many countries are also influenced by Piagetian ideas about cognitive development. The discourse – the metaphorics – here established is about how children should be allowed to be 'active', 'discover things on their own', 'work laboratively' and 'be guided by their own curiosity', they were to 'understand' and not merely 'learn by rote'. Adult intervention in children's activities and traditional teaching were seen as disturbing elements that counteract children's 'spontaneous' activities and 'independent' development. Verbal instructions – as traditional teaching was presumed to premise – were put against what was described as 'concrete' and 'self-guided activity' where the child on his or her own 'explored' the world. (Säljö, 2000, p. 58, our translation)

In an important text, written at the end of his career, Piaget himself made clear how the participation of a teacher was not seen as facilitating the child's development, rather the opposite. In the text, "Piaget's Theory", published in 1970, he wrote that "each time one prematurely teaches a child something he could have discovered for himself, that child is kept from inventing it and consequently from understanding it completely" (Piaget, 1970, p. 715). Commenting on this quote, Säljö (2000, p. 58f.) writes that this "can be seen as something of the first premise of 'child-centered' pedagogy – the child should guide his or her own development" (our translation).

While the child can certainly discover many features of his or her surrounding world through physical manipulation and observation, as suggested by Piaget, many other forms of knowledge cannot really be acquired in this way. To give an example from science education; the child can discover that objects dropped tend to fall to the ground, although some objects instead rise to the skies. However, it is difficult to see how the child through these acts of manipulation and observation could arrive at *the scientific explanation of why* this happens in one or the other way. The latter is a discursive form of knowledge that is not really there to be seen, discovered, by the child him- or herself. Rather mediation, that is, the linguistic informing of the child's experiences by a more competent partner (Pramling Samuelsson & Pramling, 2011) seems necessary for the development of this latter kind of knowledge. This claim is not specific for young children's learning, even if that is our concern in the present book. In fact this very difference and the importance of such discursive mediation can be illustrated by an empirical study of science class with older children:

Säljö and Bergqvist (1997) studied science education in the form of a physics laboratory with secondary school students (aged 13–14). The purpose of the activities followed was for the students to "acquire, by means of what is referred to as concrete experimentation, models of understanding the properties and behavior of light" (p. 393). During the laboration, the students are working on a so-called optical bench (consisting of a bench with a light source, an object such as a pen or a prism, and a screen). The following is one snippet of the ensuing conversation among some of the students and their teacher:

Anita:	It's no fun Anders [the teacher]. Nothing's happening! Nothing's happening here. Either we're stupid or it's...
ANDERS:	What are you doing then?
Eva:	Nothing.
Inga:	Nothing.
ANDERS:	I see. You're doing nothing. Well, then nothing will happen.
Eva:	Oh yes! We're doing lots of things. Yes, indeed, we're doing lots of things but still nothing's happening.
Inga:	We're finding masses of these things to do and...
Eva:	We do not know what it's for! (Säljö & Bergqvist, 1997, p. 395f.)

It is not clear to the students, even though they conduct the laboration right, what they are expected to see and why this is relevant. As Säljö and Bergqvist extensively argue, what the students are expected to see is not really there to be seen. The laboration is only illustrative if seen in terms of certain institutional concepts of physics that the teacher sees through but the students do not. For example, that there is a shadow on the screen behind a pen does not for the students become an instance of some properties of light (that it cannot go through such solid objects). Being able to see the laboration as illustrative of the properties of light requires certain sociocultural experiences that are typically appropriated through participation in schooling. However, if the students are expected to discover these by themselves, they are not supported properly in becoming members of that scientific knowledge. In other terms, the discrepancy between the expectation and the outcome as evident in the students' response makes clear the important difference between what in the language of theory of science would be referred to as 'induction' and what in cultural-historical theory is referred to as 'mediated action'. In a similar vein, Fleer (2009) has shown how without teacher and children being coordinated in perspectives – sharing semiotic mediation – they will engage in parallel, disjoint activities.

11.3 Children's Interest and How to Nurture It

One purpose of introducing children to elementary science may be to make children interested in or, if they already are so, build upon their interest in nature and how its processes may be understood. In the pedagogical literature there has been a long-lasting interest in what kinds of questions teachers pose to children (e.g., Cazden, 2001; Siraj-Blatchford & Manni, 2008; Wells, 1999; Wood, 1998). However, whether children themselves raise questions, and if so, what kinds of questions, and how these questions are responded to, have not been given the same attention. A recent study has investigated precisely these matters in the context of early childhood science education. In an empirical study of children's questions, Thulin (2010) analysed data from a sustained theme work on 'soil' in a Swedish preschool. A group of 12 children and their 3 teachers were followed with a video camera. Investigating the entire transcript of these learning events taking place over a 2-month period, she asked: (1) What do the children ask questions about? and (2) Can any developmental trend be discerned in the children's questions during the course of the theme? Hence, the first research question concerns what is thematised in the questions the children ask and the second research question concerns whether children's questions change over time. Summarising the findings in relation to the first research question, Thulin reports that the children's questions can be categorized under three headings: Questions about the content/the topic (soil, what it is, processes of decomposition etc.), Questions about the tools (e.g., magnifying glass), and Questions outside the theme (e.g., asking where an absent child is). The two first categories also have several sub-heading that we will not discuss here. In addition to categorizing the children's questions, Thulin (2010) also makes a simple quantification of these. Of the in total 206 questions asked by the children during the theme, the number of questions within each category is: 173 (Questions about the content/topic), 22 (Questions about the tools), and 11 (Questions outside the theme). One conclusion from this is that the children do ask questions and that these thus warrant analysis in, and consideration to, their science learning. A second conclusion is that the children are obviously greatly interested in the topic, the theme, since the great majority of their questions are directed towards finding something out within this. A question for education is of course how this interest in these young children could be nurtured and cultivated throughout their education.

Analysing the quantitative data on the questions posed by the children, two other important and interesting findings are reported. First, the number of questions asked by the children increases throughout the duration of the theme. On the first occasion, the children ask merely six questions. On each of the two final occasions, the children ask 48 questions. Hence, the children ask far more questions at the end than at the beginning of the theme work. This may be somewhat contrary to common sense, assuming that the less one knows the more questions one may ask. However, this result indicates the opposite. That is, in order to be able to ask (relevant) questions, the children need to gain some experience of a domain before they can ask questions to learn more. Second, not only the number of questions asked by the children changes but also the kinds of questions they ask. As already mentioned, the most common kind of question was questions about the content/the topic. However, this is also the kind of question that increases during the course of the theme. On the first occasion, four questions have this focus, while on the last occasion all 48 questions asked by the children are of this kind. Hence, not only are the children focused on, interested in finding out about, the topic, they also become more so the more experiences they gain of this topic. While the findings to the two research questions are certainly encouraging to educators, and stand in rather sharp contrast to frequently expressed fears of children not being interested in science learning in their later schooling, there is a third issue that should be considered: How do the teachers respond to the children's questions? This issue was not analysed within the framework of Thulin's study. However, in her work she hints at the teacher often responding to the children's questions by posing a new question or simply repeating the child's questions. This may be due to the teachers nurturing an ideal for early education as supposed to be guided by children exploring and themselves finding out things, or it may be due to the teachers, as generalists, not being knowledgeable enough in this particular domain to answer the children's questions. For research, studying systematically how teachers do respond to children's questions is pressing. As we have argued in this book how children's experiences are responded to by, for example, a teacher is decisive for what developmental challenges, opportunities, and support they encounter.

As was shown in this chapter, an encounter is an educational encounter under specific conditions. Similarly, a scientific encounter needs a context, as was shown in Chap. 2, which supports children to notice and use the science that is afforded through their social and material environment. Only then, is the encounter scientific.

References

Bergqvist, K. (1990). *Doing schoolwork: Task premises and joint activity in the comprehensive classroom* (Linköping studies in arts and science, 55). Linköping, Sweden: Linköping University.

Cazden, C. B. (2001). *Classroom discourse: The language of teaching and learning*. Portsmouth, NH: Heinemann.

Edwards, D., & Mercer, N. (1987). *Common knowledge: The development of understanding in the classroom*. London: Methuen.

Fleer, M. (2009). Supporting scientific conceptual consciousness or learning in 'a roundabout way' in play-based contexts. *International Journal of Science Education, 31*(8), 1069–1089.

Ivarsson, J. (2003). Kids in Zen: Computer-supported learning environments and illusory intersubjectivity. *Education, Communication & Information, 3*(3), 383–402.

Lave, J., & Wenger, E. (1991). *Situated learning: Legitimate peripheral participation*. Cambridge, UK: Cambridge University Press.

Mercer, N. (1995). *The guided construction of knowledge: Talk amongst teachers and learners*. Clevedon, UK: Multilingual Matters.

Mercer, N. (2000). *Words and minds: How we use language to think together*. London: Routledge.

Mercer, N. (2008). Talk and the development of reasoning and understanding. *Human Development, 51*(1), 90–100.

Mercer, N., & Littleton, K. (2007). *Dialogue and the development of children's thinking: A sociocultural approach*. London: Routledge.

Middleton, D., & Edwards, D. (Eds.). (1990). *Collective remembering*. London: Sage.

Nilholm, C., & Säljö, R. (1996). Co-action, situated definitions and sociocultural experience: An empirical study of problem-solving in mother-child interaction. *Learning and Instruction, 6*(4), 325–344.

Piaget, J. (1970). Piaget's theory. In P. H. Mussen (Ed.), *Carmichael's manual of child psychology* (3rd ed., Vol. 1, pp. 703–732). New York: Wiley.

Pramling, I. (1990). *Learning to learn: A study of Swedish preschool children*. New York: Springer.

Pramling, I. (1994). *Kunnandets grunder: Prövning av en fenomenografisk ansats till att utveckla barns sätt att uppfatta sin omvärld* [The foundations of knowing: Outlining a phenomenographic approach for developing children's ways of understanding their world] (Göteborg studies in educational sciences, 94). Göteborg, Sweden: Acta Universitatis Gothoburgensis.

Pramling, I. (1996). Understanding and empowering the child as a learner. In D. R. Olson & N. Torrance (Eds.), *The handbook of education and human development: New models of learning, teaching and schooling* (pp. 565–592). Cambridge, MA: Blackwell.

Pramling Samuelsson, I., & Asplund Carlsson, M. (2003). *Det lekande lärande barnet i en utvecklingspedagogisk teori* [The playing-learning child in a developmental pedagogy theory]. Stockholm: Liber.

Pramling Samuelsson, I., & Pramling, N. (2011). Didactics in early childhood education: Reflections on the volume. In N. Pramling & I. Pramling Samuelsson (Eds.), *Educational encounters: Nordic studies in early childhood didactics* (pp. 243–256). Dordrecht, The Netherlands: Springer.

Rogoff, B. (1990). *Apprenticeship in thinking: Cognitive development in social context*. New York: Oxford University Press.

Rogoff, B. (1995). Observing sociocultural activity on three planes: Participatory appropriation, guided participation and apprenticeship. In J. V. Wertsch, P. Del Rio, & A. Alvarez (Eds.), *Sociocultural studies of mind* (pp. 139–164). Cambridge, MA: Cambridge University Press.

Rogoff, B. (2003). *The cultural nature of human development*. Oxford, UK: Oxford University Press.

Rommetveit, R. (1974). *On message structure: A framework for the study of language and communication*. London: Wiley.

Säljö, R. (2000). *Lärande i praktiken: Ett sociokulturellt perspektiv* [Learning in practice: A sociocultural perspective]. Stockholm: Prisma.

Säljö, R. (2005). *Lärande och kulturella redskap: Om lärprocesser och det kollektiva minnet* [Learning and cultural tools: On processes of learning and collective memory]. Stockholm: Norstedts.

Säljö, R., & Bergqvist, K. (1997). Seeing the light: Discourse and practice in the optics lab. In L. B. Resnick, R. Säljö, C. Pontecorvo, & B. Burge (Eds.), *Discourse, tools, and reasoning: Essays on situated cognition* (pp. 385–405). Berlin, Germany: Springer.

Säljö, R., Riesbeck, E., & Wyndhamn, J. (2001). Samtal, samarbete och samsyn: En studie av koordination av perspektiv i klassrumskommunikation [Conversation, collaboration and intersubjectivity: A study of coordination of perspectives in classroom communication]. In O. Dysthe (Ed.), *Dialog, samspel og læring* [Dialogue, interaction and learning] (pp. 219–240). Oslo, Norway: Abstrakt.

Siraj-Blatchford, I. (1999). Early childhood pedagogy: Practices, principles and research. In P. Mortimer (Ed.), *Understanding pedagogy and its impact on learning* (pp. 20–45). London: Paul Chapman.

Siraj-Blatchford, I. (2007). Creativity, communication and collaboration: The identification of pedagogic progression in sustained shared thinking. *Asia-Pacific Journal of Research in Early Childhood Education, 1*(2), 3–23.

Siraj-Blatchford, I., & Manni, L. (2008). 'Would you like to tidy up now?' An analysis of adult questioning in the English Foundation Stage. *Early Years, 28*(1), 5–22.

Stone, C. A. (1998). The metaphor of scaffolding: Its utility for the field of learning disabilities. *Journal of Learning Disabilities, 31*(4), 344–364.

Sun, J., & Rao, N. (2012). Scaffolding preschool children's problem solving: A comparison between Chinese mothers and teachers across multiple tasks. *Journal of Early Childhood Research, 10*(3), 246–266.

Sylva, K., Melhuish, E., Sammons, P., Siraj-Blatchford, I., & Taggart, B. (Eds.). (2010). *Early childhood matters: Evidence from the effective pre-school and primary education project.* London: Routledge.

Thulin, S. (2010). Barns frågor under en naturvetenskaplig aktivitet i förskolan [Children's questions during a science activity in preschool]. *Nordisk Barnehageforskning, 3*(1), 27–40.

Tomasello, M. (1999). *The cultural origins of human cognition.* Cambridge, MA: Harvard University Press.

van Oers, B. (1998). From context to contextualizing. *Learning and Instruction, 8*(6), 473–488.

Vygotsky, L. S. (1987). *The collected works of L. S. Vygotsky, Volume 1: Problems of general psychology, including the volume* Thinking and Speech (R. W. Rieber & A. S. Carton, Eds., N. Minick, Trans.). New York: Plenum.

Wells, G. (1999). *Dialogic inquiry: Towards a sociocultural practice and theory of education.* New York: Cambridge University Press.

Wells, G. (2007). Semiotic mediation, dialogue and the construction of knowledge. *Human Development, 50*(5), 244–274.

Wertsch, J. V. (1979). From social interaction to higher psychological processes: A clarification and application of Vygotsky's theory. *Human Development, 22*, 1–22.

Wertsch, J. V. (1998). *Mind as action.* New York: Oxford University Press.

Wertsch, J. V. (2007). Mediation. In H. Daniels, M. Cole, & J. V. Wertsch (Eds.), *The Cambridge companion to Vygotsky* (pp. 178–192). New York: Cambridge University Press.

Wood, D. (1998). *How children think and learn* (2nd ed.). Oxford, UK: Blackwell.

Wood, D., Bruner, J. S., & Ross, G. (1976). The role of tutoring in problem solving. *Journal of Child Psychology and Psychiatry, 17*(2), 89–100.

Chapter 12
Theoretical and Conceptual Insights – Representations in Science

Niklas Pramling

Abstract This final chapter for the third part of the book illustrates the key points made in the chapters of this book and discusses these within the framework of cultural-historical theory. Semiotic mediation of cultural tools and practices in the differing forms feature in this section. In science, and science education, mediation is more commonly referred to in terms of 'representation'. Taking a cultural-historical perspective, cultural tools such as graphic data or speech, do not simply re-present phenomena and processes, they also constitute these in distinct manners for various purposes in different activities. Different mediation not only 'map' different aspects of something, but mediation contributes to how we conceive this 'something' to be.

Keywords Semiotic mediation • Representation • Cultural tools

In this chapter, we will point out some key points of the chapters in this section of the book and discuss these within the framework of cultural-historical theory. In different ways, these chapters all concern the issue of semiotic mediation (Wertsch, 2007) of cultural tools and practices. In science, and science education, mediation is more commonly referred to in terms of 'representation'. Taking a cultural-historical perspective, cultural tools such as graphic data or speech, do not simply re-present phenomena and processes, they also constitute these in distinct manners for various purposes in different activities. Different mediation do not only 'map' different aspects of something, but contribute to what we conceive this 'something' to be. A simple but effective illustration of this theoretical notion is given by Säljö (2000):

> If we think about a simple object such as an ordinary stone, it may appear simple enough to define and describe this object; we can weigh it, measure it, describe its colours and so on, and in these ways make, as it appears, an entirely exhaustive and 'objective' description of the stone. Thus, the problem of referring would be solved once and for all. However, in a sociocultural [aka cultural-historical] perspective, it is obvious that such a description of the stone, no matter how thorough, would still not embrace how the object is apprehended in different human activities and social practices. What is interesting about the stone varies between different human activities and we use it in different ways. (Säljö, 2000, p. 92, our translation)

M. Fleer, N. Pramling, *A Cultural-Historical Study of Children Learning Science*, Cultural Studies of Science Education 11, DOI 10.1007/978-94-017-9370-4_12

He goes on to illustrate his reasoning in the following manner: A stone on a lawn could be used as a goal post during a football (soccer) match; if we need to hammer a nail and do not have a hammer, we could use a stone; it could be used as an object to throw in a fight, as ornament or as an object with extra-human (deity) importance (Säljö, 2000). These, and many other possible uses of a stone, constitute it in different activities as different kinds of objects. These are not merely different aspects of the same stone; in many of these cases, the object is not – within its situated activity – a stone (but a goal post, a weapon etc.). This example may appear to be a long way from the theme of the present book, children learning science. However, familiarizing oneself with a new domain of knowing, among other things, means to learn to constitute phenomena in new and often unfamiliar ways, for example, to see features of animals as characteristics of evolutionary processes or see a cat as a predator (rather than as a pet or simply an animal). New ways of constituting phenomena and processes, as characteristic of scientific knowledge, means to conceive of these in terms of a new set of concepts. Concepts are in a sense decontextualized from here-and-now; they carry meaning over and beyond particular instances.

Developing conceptual understanding in a strict sense is much demanding of the learner (child and adult alike) and probably much rarer than we may think. Vygotsky (1998) uses the distinction between 'pseudo concepts' and 'concepts (proper)', arguing that the former means to generalize on the same level of abstraction, for example, being able to give additional examples of animals, without mastering the concept in a strict sense, that is, being able to clarify what an animal is. Even as adults, we can often give additional examples of something, for example, sports (swimming, football, slalom etc.) without being able to define what a sport is (encompassing different sports and distinguishing sport from game; cf. Wittgenstein, 1953, for an interesting analysis of this issue). Vygtosky further argues that pseudo concepts are important meeting places for child and adult; with these terms (pseudo concepts), interlocutors can talk about something without the need to share concepts in a strict sense. Pseudo concepts are therefore, he suggest, important to conceptual development. Education becomes an issue of managing the complex relation between pseudo concepts and concepts. Our concepts are likely to be pseudo concepts, while proper concepts are typical of scientific work and schooled discourse. Pseudo concepts and concepts (proper) constitute similarity amidst differences, and learning to see something as an example of something more encompassing – that is, discerning a pattern – is key to early childhood science education (cf. Björklund & Pramling, 2014). As Kress, Jewitt, Ogborn, and Tsatsarelis (2001) reason about science education, more generally:

> A central issue in learning and teaching abstractions such as 'energy' (or 'force', etc.) is seeing different particular things as similar. For example, first seeing burning wood in a fire as 'like' burning petrol in an engine and then seeing both as 'like' digesting ('burning') food. (p. 127)

One recurrent observation is that learners and teachers tend to speak metaphorically when encountering what is unfamiliar or difficult to make sense of and communicate about. We have already discussed and illustrated this feature, but a few additional points could be made. Using metaphorical or figurative terms, that is, speaking about the novel in terms of something more familiar, means to simultaneously relate and distance as integral to scientific reasoning and understanding (Kress et al., 2001). Phrased differently – and in terms of a traditional distinction in research on children's thinking and development, 'concrete' and 'abstract' – developing an understanding is not a unidirectional process from concrete to abstract. Rather, in children's sense-making practices in the form of metaphorical reasoning, they simultaneously 'concretize' (make concrete through speaking in more familiar terms) and 'abstract' (since perceiving a metaphorical relationship is an act of abstraction in being able to see some kind of similarity or analogy between diverse instances). In fact, there is an abstraction 'in-built' into cultural tools; the word 'house', for example, does not denote a particular house, but a category (Sapir, 1921; Vygotsky, 1987). Using such tools in different activities therefore also includes what Billig (1996) refers to as 'particularization', making the tool (concept) relevant for one's current concern and particular instances.

Integral to learning science is to see something as an example of something more general, to see something as an instance of a principle. Theoretical concepts constitute particular relations between situations. "One of the advantages of theoretical concepts", as contrast to more local (i.e., deictic) forms of referencing (see Chap. 11), Ivarsson (2003, p. 398) argues, "is that they, in their capacity as linguistic tools, can be used in different contexts with some meaning preserved. Or put more correctly, since they maintain a relation to earlier contexts, the meaning of concepts can more easily be recreated in new situations, a process sometimes referred to as recontextualisation (van Oers, 1998)". Appropriating cultural tools in the form of scientific concepts thus allows the learner to perceive what is observed as instances of more general and theoretically motivated phenomena or processes. Learning to see what terms are relevant and functional to speak about and perceive nature in, in itself constitute a feature of science learning. Without some communicative coordination with a teacher (or more experienced peer), the child will make sense of nature in whatever familiar terms he or she deems relevant (Fleer, 2009; see also Chap. 3). In the next chapter, we bring together these ideas and more to conclude the book. By drawing together all the themes discussed throughout the book, we present a cultural-historical model of early childhood science education.

References

Billig, M. (1996). *Arguing and thinking: A rhetorical approach to social psychology* (2nd ed.). Cambridge, UK: Cambridge University Press.

Björklund, C., & Pramling, N. (2014). Pattern discernment and pseudo-conceptual development in early childhood mathematics education. *International Journal of Early Years Education, 22*(1), 89–104.

Fleer, M. (2009). Supporting scientific conceptual consciousness or learning in a 'roundabout way' in play-based contexts. *International Journal of Science Education, 31*(8), 1069–1089.

Ivarsson, J. (2003). Kids in Zen: Computer-supported learning environments and illusory intersubjectivity. *Education, Communication & Information, 3*(3), 383–402.
Kress, G., Jewitt, C., Ogborn, J., & Tsatsarelis, C. (2001). *Multimodal teaching and learning: The rhetorics of the science classroom.* London: Continuum.
Säljö, R. (2000). *Lärande i praktiken: Ett sociokulturellt perspektiv* [Learning in practice: A sociocultural perspective]. Stockholm: Prisma.
Sapir, E. (1921). *Language: An introduction to the study of speech.* New York: Harcourt, Brace.
van Oers, B. (1998). From context to contextualizing. *Learning and Instruction, 8*(6), 473–488.
Vygotsky, L. S. (1987). *The collected works of L. S. Vygotsky, Volume 1: Problems of general psychology, including the volume* Thinking and Speech (R. W. Rieber & A. S. Carton, Eds., & N. Minick, Trans.). New York: Plenum.
Vygotsky, L. S. (1998). *The collected works of L. S. Vygotsky, Volume 5: Child psychology* (R. W. Rieber, Ed., M. J. Hall, Trans.). New York: Plenum.
Wertsch, J. V. (2007). Mediation. In H. Daniels, M. Cole, & J. V. Wertsch (Eds.), *The Cambridge companion to Vygotsky* (pp. 178–192). New York: Cambridge University Press.
Wittgenstein, L. (1953). *Philosophische untersuchungen/Philosophical investigations* (G. E. M. Anscombe, Trans.). Oxford: Basil Blackwell.

Part IV
Models of Early Childhood Teaching

Chapter 13
A Cultural-Historical Model of Early Childhood Science Education

Marilyn Fleer

Abstract The final chapter of this book brings together the central ideas discussed in previous chapters and presents a model of early childhood science education. This chapter theorises early childhood science education as a dialectical process between everyday and scientific concepts. The central concepts of the social situation of development, the relations between the ideal and the real form, imagination and creativity, and perezhivanie are reviewed in this chapter. Similarly the concepts of shared sustained thinking, rising to the concrete, intersubjectivity, mediation, metaphorical speech, anthropomorphic speech, the gap between the familiar and unfamiliar, simile, and metaphor are also revisited. The nature of institutional practices, the relations between 'telling' and 'explaining', the aesthetics of perception, and how phenomenon are culturally and socially constructed, and what this means for the role of the early childhood teacher are discussed. Further, the nature of children's drawings in science and how this contributes to children's scientific thinking and conceptual development of science concepts are considered. Together, these concepts give a different view of research in early childhood science education to previous reviews. An example of a cultural-historical model of early childhood science education in action completes the book.

Keywords Dialectical process • Social situation of development • Zone of proximal development • Imagination and creativity • Intersubjectivity, mediation, metaphorical speech, anthropomorphic speech • Ideal and the real form • Perezhivanie

13.1 Introduction

In this book we have argued that the *historical legacy of science education research* is rich but grounded predominantly in one theoretical construction of reality. Here we have found problematic the dualism between traditional concepts of what was known as *Children's Science* and contemporary perspectives on conceptual scientific development in the context of *socioscientific* pedagogies. *Children's science* or *alternative conceptions theory* only ever gave one side of the coin. A socioscientific focus on research has re-introduced the role of the teacher in determining learning and development in science in a more dialogical way. However, it was noted in our review of the literature that this research had been directed primarily towards upper

M. Fleer, N. Pramling, *A Cultural-Historical Study of Children Learning Science*, Cultural Studies of Science Education 11, DOI 10.1007/978-94-017-9370-4_13

primary and secondary pupils where *argumentation* on socioscientific ideas features. This latter pedagogical approach is clearly not possible with preschool children. What has been missing for the early childhood field has been a study of the dialectical relations between children's thinking and the social and material conditions which develop *curiosity* and an *affective* reading of contexts and concepts in science (Part I), and a deeper understanding of children's *scientific representation* and how they are culturally constructed and socially mediated by teachers in early childhood classrooms and centres (Part III).

We also noted that the dualism between everyday alternative conceptions of the world and scientific constructions of the world as legitimised by scientists was unhelpful for progressing a model of science learning for very young children. We found more fruitful the idea that everyday conceptions in science were *integral* rather than being conceptualised as getting in the way of learning science concepts. Here we drew upon Vygotsky's theorisation of everyday concepts and scientific concepts in order to put forward a *dialectical view of the relations between everyday and scientific concepts when learning science.* In this sense we have returned to a focus on the child and their scientific thinking, but in relation to the social and material conditions in which that thinking is taking place. It is this *dialectical relations* that has been the focus of our research attention. This dialetctial relation has also been noted in recent research by Roth, Goulart, and Plakitsi (2013) who argue for a *dialectic of participation.* We observed that this dialectical view of science learning is generally at odds with the dominant research attention on early childhood science education research in many countries around the world, which we reviewed in the second part of this book. Our position was featured through the dialectical concepts of imagination and creativity, everyday and scientific concept formation, ideal and real, and the social situation of the environment. A natural tension exists between each of these concepts, and it is this tension that provides the movement in learning and development of children in science.

In this book we also discussed the idea that science knowledge is not static. In the first two parts of the book we put forward the view that scientific concepts change over time and across communities. We argued that how these understandings are formed and researched, also varies and evolves across cultural communities. A universal view does not take account of what children and researchers bring to science education, or how this shapes how knowledge is formed, or indeed what forms of knowledge are valued – empirical, narrative or theoretical (see Chap. 5).

In the third part of the book we discussed how children respond to, encounter and represent their scientific understandings. Specifically, we argued against a process of osmosis of science learning, as has been the pedagogical fashion in early childhood science education across many communities (e.g. developmentally appropriate practice or discovery learning). We drew upon a range of contemporary pedagogical approaches theorised and researched from a cultural-historical perspective. This section of the book, combined with the first section of the book, made visible some important cultural-historical concepts that better informed our understanding of scientific conceptual development of early childhood children than previous theories – constructivism and developmentally appropriate practice.

Together these three parts of the book theorise early childhood science education as a dialectical process between everyday and scientific concepts. To make our case, we specifically worked with a system of concepts: the social situation of development, the relations between the ideal and the real, imagination and creativity, and perezhivanie. We also discussed the concepts of shared sustained thinking, rising to the concrete, intersubjectivity, mediation, metaphorical speech, anthropomorphic speech, the gap between the familiar and unfamiliar, simile, and metaphor. Here we specifically examined the nature of institutional practices, the relations between 'telling' and 'explaining', the aesthetics of perception, and how phenomenon are culturally and socially constructed, and what this means for the role of the early childhood teacher. We also examined the nature of children's drawings in science and how this contributes to children's scientific thinking and conceptual development of science concepts. Together, these concepts give a different view of research in early childhood science education to previous reviews, such as that offered by Eshach (2006), Martin, Jean-Sigur, and Schmidt (2005), Metz (2006) or that which dominates much of content edited by Saracho and Spodek (2008). They are in line with Roth et al. (2013) who also draw upon cultural-historical concepts for discussing science education during early childhood.

In this final chapter in the book we bring together these central ideas and present a model of early childhood science education. We believe our review and theorization offers a development in thinking that is productive for both research methodologies and pedagogies in early childhood science education.

13.2 A Cultural-Historical Informed Pedagogical Model of Early Childhood Science Education

In this chapter we draw on both the concepts reviewed and empirical content discussed in this book to introduce our pedagogical model for creating the conditions for science learning in preschool settings. We know from research that the physical preschool environment affords many possibilities for science learning in play-based settings (see Chap. 2). However, as has been shown throughout this book, it is the relations between the child and the environment through the teacher that provides the best opportunities for maximising the learning of science. As such, we need to conceptualise the development of science concepts in relation to learning. In order to do this, we must first think about the relations between learning and development, and second, we must conceptualise these relations in practice as a pedagogical model suitable for young children in play-based contexts. What is unique about early childhood science education is not just the nature of the preschool child, but also the play-based environment in which the child learns science and develops as a human being. Our pedagogical model (see both Figs. 13.1 and 13.2) is specific for early childhood children and the play-based contexts in which they are taught science.

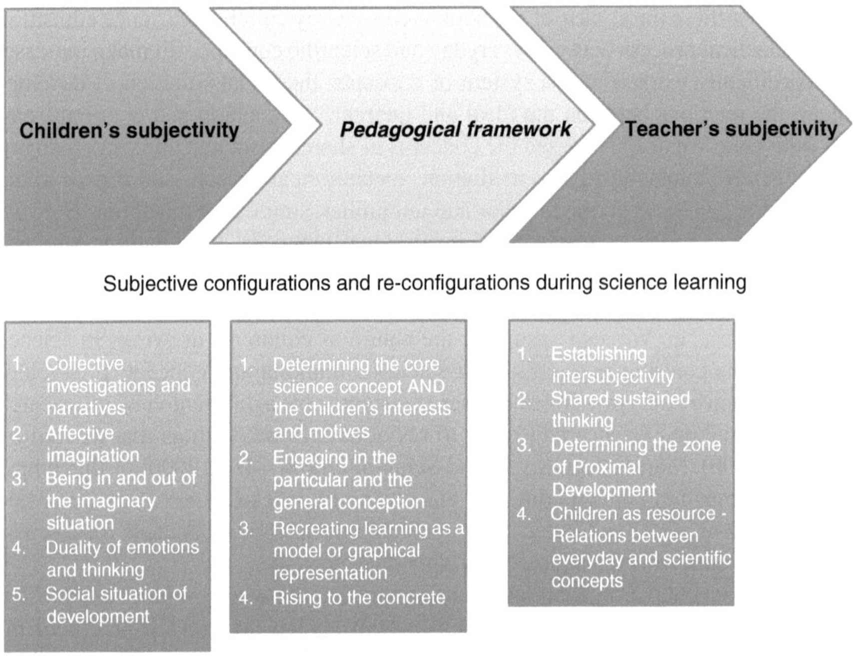

Fig. 13.1 A cultural-historical model of early childhood science education

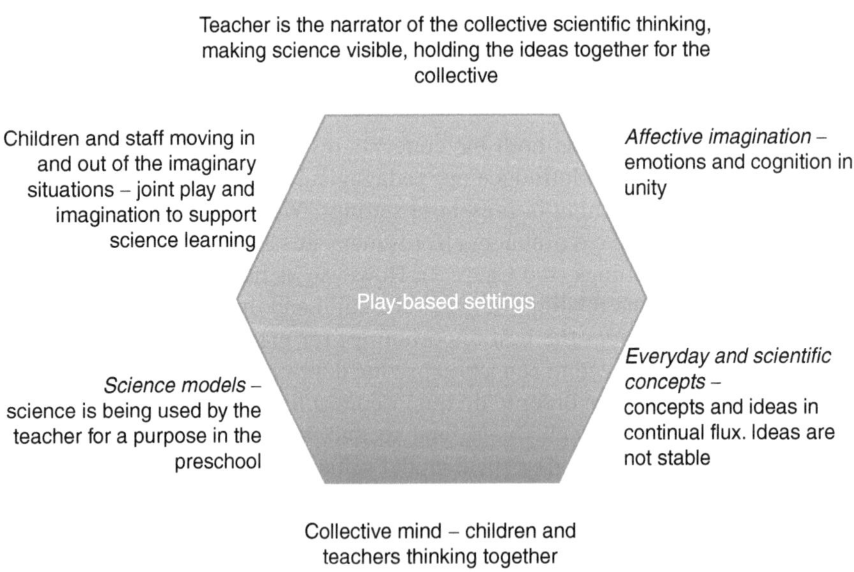

Fig. 13.2 Core elements of what is unique about early childhood science teaching and learning

13.2.1 The Relations Between Learning and Development

Learning is defined from a cultural-historical perspective as a change in the child's "relation to another person and activities in specific settings" (Hedegaard & Fleer, 2013, p. 183) as we see when a child learns scientific concepts. As a result of learning, children begin to act differently, because they have new insights into how their world works. In contrast, a cultural-historical conception of development can be understood as a process where "children's motive orientation and engagement in different activity settings change qualitatively" and as such their leading motive changes (Hedegaard & Fleer, 2013, p. 183). For example, we see this qualitative change in the person when s/he no longer wish to actively play all of the time, but rather has a new orientation to learning, and wishes to engage in the serious study of how her/his world works, or in learning how to read and write. Here the child's orientation has changed from play to learning. Vygotsky (1998) argued that learning has a huge impact upon children's overall development. Learning of concepts is a cultural practice, and as such contributes to the cultural development of the child. Hence, culture helps explain the relation between learning and the development of a child. Whilst biology is important, a cultural-historical view would suggest that it is not the driving force of children's development.

The child who learns concepts, begins to think and act in new ways, according to his or her new understanding of the world around him or her. With new scientific understandings about his or her world, the child can act differently and through this afford new possibilities and learning. A further analogy of this idea, is the child who understands the pointing gesture. Pointing as a cultural and not biological phenomenon, means that children who understand this cultural practice are able to form a different type of relationship to their environment and to others. For instance, they can direct people's attention to something, they can signal the need for something. This is possible because the child has a new understanding, an understanding which acts as a cultural tool, directed to another human being, changing the child's relationship from direct interaction with the environment (i.e. getting the object themselves) to interacting with the environment through another person (pointing to the object for another person to pay attention to or to retrieve for them). The environment does not change, but rather it is a cultural change in the child which affords a new way of interacting *with that same* environment. The child is subjectively configured simultaneously to a social and material world. This process of development can be maximised in preschool centres when a robust pedagogical framework for science learning is used by the teacher.

Through learning scientific concepts children gain a new sense of the situation, and they in turn think about their world in new ways. Over time, and through the learning of many new concepts, we begin to see a qualitative change in the child's development. We can use the metaphor of the tadpole and the frog to explain this qualitative change in development. The tadpole is not a miniature frog, but rather a qualitatively different physically represented organism to the frog. Qualitative change is not an incremental change where a small frog becomes a big frog, but rather it is a qualitative change from a tadpole to a frog. Children's development is a qualitative change. Learning progressively contributes to this qualitative change of the whole child.

13.2.2 A Pedagogical Framework for Scientific Learning

In Fig. 13.1 we bring together the children's perspective of science learning, a conceptualisation of the teacher of science, and a pedagogical framework for creating the conditions for science learning. Each of these elements contributes to the collective subjectivity and the subjective sense (Gonzelez Rey, 2012) that children make of scientific encounters (see Chap. 10). It is acknowledged that these three elements are totally interrelated, and constitute the collective conditions for learning science, despite the illusion of their separation in the model. The content presented in the model should be viewed as a system of concepts that together speak to the social situation of science learning and through this contribute towards the development of children as a whole.

In Fig. 13.1 we conceptualise the children's and the teacher's subjectivity during science learning as a creative and imaginative *production* (Gonzelez Rey, 2009), where a new scientific *sense* is being *configured* and reconfigured (Gonzelez Rey, 2012). In drawing upon the work of Gonzelez Rey (2009), we can theorise this emotional and cognitive flux as a *subjective configuration and reconfiguration*. We know from our research (see Chap. 5) and from the research process itself with young children (see Chap. 7), that children's thinking in science is emotionally charged (see Chap. 3), and continually changes within moments (see Chap. 1). That is, we should not view children's thinking in science as static (see Chap. 1), but rather as in constant motion (see examples in Part III). We capture this dynamic flux in the pedagogical model shown below, through foregrounding children's subjectivity (solid arrow 1) and the teacher's subjectivity (solid arrow 3) throughout all educational encounters that support science learning.

What is key here is embracing children's dynamic thinking, always in flux, always emotionally charged, and always connected to those around them. Curiosity is constructed, enacted, and learned when experienced through a rich but teacher engaged process for science learning. That is, the teacher has an active role in the process (see Part III) and children are not left to discover the science in the situation by themselves, as has dominated early childhood education, where an individualistic construction of learning if featured. We showed the active role of the teacher in Chap. 6 where the teacher contributed to framing the child's learning through investigating why the ant was in the wrong place. The teacher deliberately build theoretical knowledge for the child through the introduction of tools and resources which allowed the children to build a theoretical model of ecosystem – where the dynamic relations between insect, food source, and habitat was actively supported.

Without the teacher structuring experiences to build a theoretical model of the ecosystem of the ant, the child would have continued to look at insects randomly in the environment.

It is important for the teaching of science that the pedagogy celebrates the subjective sense of science that children bring with them to the social and material preschool environment (as a resource). It is the teacher who creates the dynamic conditions, and it is s/he who acknowledges the *thinking flux* and subjectivity of both children and self (own subjectivity – e.g. enjoyment or aversion for science). The emotional and cognitive unity continues to be configured and re-configured during the teaching and learning of science, thus contributing to the qualitative changes and hence overall development of the child. As the child learns something new, s/he interacts in new ways within the learning situation. This affords new conceptual possibilities about how the child's social and material environment is understood, but it also can change how the child feels about him or herself as a learner of science concepts (Chap. 3). We also saw this in the examples given in Chaps. 2 and 3 where the teacher supported the children's scientific thinking through very pleasurable experiences of creating rainbows, exploring light, and understanding the weather patterns for playing outside. The children and the teacher were positively engaged in science learning, and the learning of science was positively contributing to the quality of the children's outdoor play and general experiences.

What we noted in Part I of this book, was that science learning is a highly imaginative and creative act. That children move in and out of imaginary situations in their play, taking with them their growing conceptual understandings in science. Children are both thinking and feeling as they experience science learning (perezhevanie, see Chap. 3), and curiosity is ignited when children have an affective relationship to the content and the process of science learning.

In Chap. 3 we noted how in play based settings that science learning is affectively charged, and imagination was central for realising scientific concept formation. Children collectively develop a consciousness of scientific and technological concepts and emotionality by working together with other children to solve the problem. Children use a *scientific narrative* to collectively work together to solve scientific and technological problems. Children in their role-play of scientific narratives also *collectively* begin to anticipate the results of each others' actions in the play, begin to anticipate their own actions, including image-bearing dramatization, verbal descriptions, prop use and transformation, and importantly, the scientific solutions created through the collaborative support of the teacher. It is the border of the imaginary world and the concrete world that creates a dialectical relation and emotional tension that promotes scientific conceptual development. In scientific investigations, children's feeling state becomes connected with their learning as they anticipate *finding a solution*. Through consciously considering feeling states in science, emotions become intellectualized, generalized, and anticipatory, while cognitive processes acquire an affective dimension, performing a special role in meaning discrimination and meaning formation (e.g., gut feeling this is going to work).

But what is critical here is how teachers help children in knowing what is noteworthy to pay attention to in science learning. This is reflected in the model as:

1. *Establishing intersubjectivity.* Without teachers and children coming to know each other as social players, it is difficult for a teacher to know what will hold a child's attention, what will be meaningful, and what would be an authentic educational encounter. Similarly, without intersubjectivity, the teacher could never position children as a resource, with ideas, curiosity, questions, and interest to role-play aspects of their own social and material world.
2. *Shared sustained thinking.* Without teachers building on or stimulating engaging and deeply theorised dialogues with young children about concepts in the everyday world and concepts in science, children would not see the richness of the science in their world, or would have limited opportunities for thinking scientifically about everyday life. It is the sustained nature of the conversation with a child in play, in everyday life exchanges, and in scientific encounters, that establish and maintain a scientific attitude to life, learning and thinking.
3. *Zone of proximal development.* Teachers who use the concept of the Zone of Proximal Development can identify the actual and the potential development of children. They know that it is in the ZPD that we find the maturing processes of a child's development. The ZPD is about understanding or assessing those maturing processes that become evident when the child is working in cooperation or is guided by others (see Vygotsky, 1998). Teachers determine children's actual and potential development. We conceptualise the ZPD as a form of cooperation between the child and an adult, where the child can with support engage with and conceptualise concepts as determined by their ZPD. It is only those concept which are already within the child's psychological grasp and experience that can be realised during interaction that form the ZPD. Actual development is determined as an independent interaction, and conceptualised as the already formed functions and processes of the child. It is the relations between the actual development of the child and the ideal form of development in cooperation with another, where we see development being progressed. The teacher's role is central here for realising a productive relation between the actual and the ideal.
4. *Children as resource.* In the context of learning, we see that the concept of the ZPD directs our attention to determining the actual conceptual understanding of the child and through the active relations between the child's actual understandings and the ideal concepts, that we see a movement from everyday understandings to scientific understandings. The pedagogical framework creates the conditions whereby the child's everyday understandings act as a resource during the learning of scientific concepts. The child's experiences, motives and interests are key to the pedagogical situation, giving meaning to the educational encounter.

To build theoretical knowledge in science requires a particular kind of cooperation by the teacher with the children, so that children look with scientific eyes, as they build an understanding of their finds in relation to the knowledge system that they are encountering. In Chap. 5 we specifically examined the child's journey into this knowledge system in relation to what conditions had been created by the teacher. This is reflected in the model as:

1. *Core concepts.* Determining what might be the core scientific concepts to be learned
2. *Particular-general dialectic.* Engaging children in thinking about both the particular element (e.g., ant), and the general concept (e.g. species classification 'insect')
3. *Models.* Supporting children to re-create their learning as models and graphical representations, and using metaphors and similes
4. *Rising to the concrete.* Creating the conditions that allowed for children to conceptually *rise to the concrete* by having the opportunity to consider how the abstract knowledge (e.g., species classification) was formed in the first place (observing form, function, food source, and habitat of a particular insect) by scientists.

Taken together, the elements discussed above represent the subjective conditions that determine the social situation of development of the child during the iterative process of learning science concepts in play-based settings. Whilst the science encounters are collectively constructed, how each individual child experiences this same set of scientific encounters will depend upon what he or she bring to the that same situation. Each child will have different prior everyday experiences of their world which they s/he draws upon when making sense of scientific encounters. The scientific experience will be affectively refracted through how the child feels about the learning experience. Scientific curiosity is not just a cognitive activity but is affectively charged process. In this book we have foregrounded the unity of emotions and cognition and argued that affective imagination be a central part of a pedagogical model for teaching science. Yet as Zembylas (2008) suggests "affective factors have been largely neglected in science education research which has been dominated by "conceptual change" view of learning (Alsop and Watts, 2003)" (p. 66), and "relatively little work has explicitly addressed affect, feelings, or the emotions compared to the large literature on attitudes to school science" (p. 67). In our model we not only acknowledged the place of emotions in science, but suggest that this acts like a glue holding all the other elements of our model together as a dialectical unit. Here there exists an indivisibility of environment and the personality of the child, as a form of *perezhivanie*. Here perezhivanie is "all the personal characteristics and all the environmental characteristics ... represented in an emotional experience *[perezhivanie]*" (Vygotsky, 1994, p. 341; Original emphasis). This also means that what takes place in the preschool cannot be conceptualised without considering what takes place outside of the preschool, in the family home and in the community. Here we agree with Roth (2012) who has argued for a dialectical approach for science education where what is learned in schools be authentic and valued by the child and therefore useful for life outside of classrooms. To achieve this requires a view of early childhood science learning that brings together both the children's subjectivity and the teacher's subjectivity as a collective and dynamically interacting enterprise, as has been conceptualised by Gonzalez Rey (2012) as a subjective configuration: "The person is always within a network of symbolic processes and emotions, that characterises their social existence. Human activities and relations are configured to each other within a complex subjective system of human existence" (p. 49). It can be argued that learning science in play-based environments affords a very complex network of symbolic processes and emotions that continue to be configured and re-configured. Consequently, it becomes important to capture the essence of what is early childhood learning and teaching of science? What are the core unique features that are distinct from primary or even secondary science learning and teaching? What are the features which are unique to play-based settings and the nature of the young learner? The uniqueness is symbolically represented in Fig. 13.2 where six elements are foregrounded as the core features of the teaching-learning process of early childhood science.

Whilst Fig. 13.1 presented what mattered for learning science, Fig. 13.2 takes from the research and Fig. 13.1 those core elements that are specifically unique to learning science in play-based contexts for preschool aged children. Here affective imagination is foregrounded, but in the context of the teacher acting as the narrator of the collective scientific thinking, making science visible, holding the ideas together for the collective. We saw examples of this throughout the content of this book. Young children need support with noticing the science in everyday situations, as well as help with linking their thinking from one day to the next in preschool settings. Their ideas are not stable, and their thinking is in constant flux. Teachers can support this process through creating models and supporting children to construct representations of their growing ideas in science – as artefacts of their thinking and as cultural tools to support new thinking. Moving in and out of imaginary situations allows children to think iteratively about the concrete object and the abstract representation. Role-play as well as imaginary play supported by the teacher creates many possibilities for also thinking abstractly. We saw examples of this in Part I of the book. In essence, a play-based setting affords the need for a sense of the *collective mind* as children and teachers engage in scientific encounters where scientific ideas are iteratively explored on one day, from day to day, and over the course of weeks, and even the year. However, these unique features of the nature of young scientific learner have not been adequately recognized in science education. Figure 13.2 begins to make the uniqueness of early childhood science learning in play-based settings visible. We illustrate the model in action through a brief example shown below in Table 13.1

Table 13.1 An example of a cultural-historical model of early childhood science education in action

Key concept	Explanation	Example
Teacher as narrator	The teacher holds the scientific narrative together – in one day, over one week, and between children.	Children aged 2 for 5 years demonstrate interest in learning about their bodies as a result of someone being away sick – asking about why Isobella is not at child care. The teacher plans a range of experiences to develop their scientific thinking over a period of 4 weeks. However, to hold together the learning journey, the teacher does the following:
		Each day at group time she re-visits what the children did the previous day; She uses 'thinking books' which are A4 sheets of the children's ideas, thinking and investigations, that are collected and stapled together and read out at grouptime or to families; She also has the children sit in circle to show and tell about their learning, using their thinking books; She does group mind maps, concept maps and other posters of investigations, including storyboards and photographs, iPad animation; children's posters, as records of the ongoing activities. She references these regularly throughout the day.
Collective mind	Children and teacher are thinking together. The teacher is in the imaginary situation with the children as they imagine both play and learning with the scientific concepts.	The children and teacher create a life sized human body from boxes, fabric and plastic that they can crawl inside. The children together with the teacher, enter through the mouth of the their human body, passing through all the major organs. The children and teacher are in the imaginary situation together. The children make an enormous heart from fabric and the children enter into the imaginary circulatory system of the body, projecting out, naming different organ they take oxygen and food too.
Children moving in and out of imaginary situations	The children both imagine the abstract concepts of science and the concrete situation	The children create a Play World of the human body. That is, they enter into the fairytale of Jack and the beanstalk, and when the giant falls to the ground and is unconscious, they undertake a series of investigations/adventures, diagnostics (being doctors), and together with the teacher undertake surgery of the giant. This Play World scenario is supported by visiting a hospital to learn about different procedures that can then be used back in the Play World.

(continued)

Table 13.1 (continued)

Key concept	Explanation	Example
Science models	The teacher helps the children to build a theoretical model of science concepts	The children create their own poster about the human body. They trace around each others' bodies, and then they draw what is inside their bodies, with cut out flaps, for going more deeply into the different organs they have drawn. The children add to their human body poster by making it interactive through projections and sound effects via their iPad animations. Finally, the children with support from the teacher create their own YouTube clip explaining their poster. The links are sent to families for sharing and further discussion. The core concept of a 'system of organs' is the key feature of the theoretical knowledge supported by the teacher in building the model with the children.
Everyday concepts and scientific concepts	Everyday experiences and scientific understandings are concurrently supported by the teacher	The teacher invites the children to make an animation on an iPad using playdough, coloured cardboard, string, etc to replicate their understandings of the human body. The children think deliberately about their everyday understandings, and together with the teacher they check sources (e.g. YouTube, books, expert scientists they phone).
Affective imagination	How you feel about the learning of science concepts and how the science concepts positively contribute to living and working in everyday situations matters	Story world, the interactive poster, the YouTubes, the thinking books, and the narration by the teacher to bring all the experiences from one day to the next contribute to an emotionally charged and positive experience of learning about the human body. Featuring the children's own bodies and imagination supports affective imagination.

13.3 Conclusion

As Robbins (2012) reminds us "Currently, there are a relatively small, but growing number of science education researchers who are framing their work from a socio-cultural or cultural-historical perspective (see Fleer, 2009; Fleer, Ridgway, & Gunstone, 2006; Fleer & Robbins, 2003; Giest & Lompscher, 2003; Leach & Scott, 2003; Lemke, 2001; O'Loughin, 1992; Schoulz, Säljö, & Wyndham, 2001; Traianou, 2006)" (p. 78). In this book we have not only plotted this movement (see Fleer, 2013; Fleer & March, 2006; Goulart & Roth, 2010; Mawson, 2007; Ravanis & Bagakis, 1998; Ravanis, Christidou, & Hatzinikita, 2013; Traianou, 2006), but expanded upon this body of research to give a fuller and richer picture of what constitutes a cultural-historical study of early childhood science education.

In the context of research in early childhood science education, Robbins (2009) also states that "for many in early childhood education there is a movement towards sociocultural views on learning while science education appears largely fixed on individual views of learning" (p. 78). As was shown in this book, that whilst there is more research being done by researchers from early childhood education drawing upon cultural-historical theory, we are still seeing research into children's conceptions in science published, a focus that the general science education community has largely left behind as being unproductive as they move into argumentation and a more socioscientific approach for progressing science education. Consequently, it can be argued that early childhood science education research is full of contradictions. On the one hand it has embraced cultural-historical theory by undertaking rich and progressive research, and on the other hand it continues to undertake traditional research following what the rest of science education now view as dated – alternative conceptions theory or Children's Science. But what has changed is the number of researchers actively engaged in early childhood science education research. Ten years ago very little research was being done in this area. Now there are more studies, more researchers, and more focus on what is unique about young children's learning in science. Rather than pedagogical models that were developed on research with adolescents or models suitable for primary aged children being adopted and adapted for use with young children, the early childhood community has research to better understand the nature of the very young learner. What the early childhood community does not have is access to suitable pedagogical models developed from early childhood education research. This book seeks to contribute to the early childhood community by offering a compilation and critique of early childhood research and by putting forward a pedagogical model of learning science that foregrounds affective imagination as central for play-based settings (Fig. 13.2). Through the contents of this book, we seek to make accessible the wealth of research and pedagogical discussion on the unique attributes to learning science in play-based settings.

References

Eshach, H. (2006). *Science literacy in primary schools and pre-schools*. Dordrecht, The Netherlands: Springer.

Fleer, M. (2013). *Affective imagination* in science education: Determining the emotional nature of scientific and technological learning of young children. *Research in Science Education.* Published on line Jan 2013.

Fleer, M., & March, S. (2009). Engagement in science, engineering and technology in the early years: A cultural-historical reading. *Review of Science, Mathematics and ICT Education, 3*(1), 23–48.

Fleer, M., Ridgway, A., & Gunstone, D. (2006). The cultural-historical construction of science learning within play based contexts. *Paper presented at the Thirty-seventh annual conference of the Australasian Science Education Research Association*, Canberra, Australia.

Fleer, M., & Robbins, J. (2003). 'Hit and run' research with 'hit and miss results' in early childhood science. *Research in Science Education, 33*(4), 405–431.

Giest, H., & Lompscher, J. (2003). Formation of learning activity and theoretical thinking in science teaching. In A. Kozulin, B. Gindis, V. S. Ageyev, & S. Miller (Eds.), *Vygotsky's educational theory in cultural context* (pp. 267–288). Cambridge: Cambridge University Press.

Gonzelez Rey, F. (2009). Historical relevance of Vygotsky's work: Its significance for a new approach to the problem of subjectivity in psychology. *Outlines, 1*, 59–73.

Gonzalez Rey, F. (2012). Advancing on the concept of sense: Subjective sense and subjective configurations in human development. In M. Hedegaard, A. Edwards, & M. Fleer (Eds.), *Motives in children's development. Cultural-historical approaches* (pp. 45–62). Cambridge, UK: Cambridge University Press.

Goulart, M. I. M., & Roth, W.-M. (2010). Engaging young children in collective curriculum design. *Cultural Studies of Science Education, 5*, 533–562.

Hedegaard, M., & Fleer, M. (2013). *Play, leaning and children's development: Everyday life in families and transition to school*. New York: Cambridge University Press.

Leach, J., & Scott, P. (2003). Individual and sociocultural views of learning in science education. *Science and Education, 12*, 91–113.

Lemke, J. L. (2001). Articulating communities: Sociocultural perspectives on Science Education. *Journal of Research in Science Teaching, 38*(3), 296–316.

Martin, D. J., Jean-Sigur, R., & Schmidt, E. (2005). Process oriented inquiry – A constructivist approach to early childhood science education: Teaching teachers to do science. *Journal of Elementary Science Education, 17*(2), 13–26.

Mawson, L. (2007). Introduction: Bridging the cognitive and sociocultural approaches in research on conceptual change: Is it feasible? *Educational Psychologist, 42*(1), 1–7.

Metz, K. (2006). Rethinking what is "developmentally appropriate" from a learning progression perspective: The power and the challenge. *Review of Science, Mathematics and ICT Education, 3*(1), 5–22.

O'Loughin, M. (1992). Rethinking science education: Beyond Piagetian constructivism toward a sociocultural model of teaching and learning. *Journal of Research in Science Teaching, 2*(8), 791–820.

Ravanis, K., & Bagakis, G. (1998). Science education in kindergartens: Sociocognitive perspective. *International Journal of Early Years Education, 6*(3), 315–327.

Ravanis, R., Christidou, V., & Hatzinikita, V. (2013). Enhancing conceptual change in preschool children's representations of light: A sociocognitive approach. *Research in Science Education, 43*, 2257. doi:10.1007/s11165-013-9356-z.

Robbins, J. (2009). Analysing young children's thinking about natural phenomena: A sociocultural/cultural historical perspective. *Review of Science, Mathematics and ICT Education, 3*(1), 75–97.

Roth, W.-M. (2012). Science of learning is learning of science: Why we need a dialectical approach to science education research. *Cultural Studies of Science Education, 7*, 255–277.

Roth, W.-M., Goulart, M. I. M., & Plakitsi, K. (2013). *Science education during early childhood. A cultural-historical perspective*. Dordrecht, The Netherlands: Springer.

Saracho, O. N., & Spodek, B. (Eds.). (2008). *Contemporary perspectives on science and technology in early childhood education*. Charlotte, NC: Information Age Publishing.

Schoulz, J., Säljö, R., & Wyndham, J. (2001). Heavenly talk: Discourse, artifacts and children's understanding of elementary astronomy. *Human Development, 44*(2/3), 103–118.

Traianou, A. (2006). Teachers' adequacy of subject knowledge in primary science: Assessing constructivist approaches from a sociocultural perspective. *International Journal of Science Education, 28*(8), 827–842.

Vygotsky, L. S. (1994). The problem of the environment. In J. Valsiner & R. van der Veer (Eds.), *The Vygotsky reader* (pp. 347–348). Oxford: Blackwell.

Vygotsky, L. S. (1998). Child psychology. In *The collected works of L.S. Vygotsky, vol. 5* (L. S. Vygotsky, Ed., M. J. Hall, R. W. Rieber, Trans., ed. English translation). New York: Kluwer Academic/Plenum Publishers.

Zembylas, M. (2008). Affect and early childhood science education. In O. N. Saracho & B. Spodek (Eds.), *Contemporary perspectives on science and technology in early childhood education* (pp. 65–85). Charlotte, NC: Information Age Publishing.

MIX
Papier aus verantwortungsvollen Quellen
Paper from responsible sources
FSC® C105338
FSC
www.fsc.org

If you have any concerns about our products,
you can contact us on
ProductSafety@springernature.com

In case Publisher is established outside the EU,
the EU authorized representative is:
Springer Nature Customer Service Center GmbH
Europaplatz 3, 69115 Heidelberg, Germany

Printed by Libri Plureos GmbH
in Hamburg, Germany